CAMBRIDGE LIBRARY COLLECTION

Books of enduring scholarly value

Earth Sciences

In the nineteenth century, geology emerged as a distinct academic discipline. It pointed the way towards the theory of evolution, as scientists including Gideon Mantell, Adam Sedgwick, Charles Lyell and Roderick Murchison began to use the evidence of minerals, rock formations and fossils to demonstrate that the earth was older by millions of years than the conventional, Bible-based wisdom had supposed. They argued convincingly that the climate, flora and fauna of the distant past could be deduced from geological evidence. Volcanic activity, the formation of mountains, and the action of glaciers and rivers, tides and ocean currents also became better understood. This series includes landmark publications by pioneers of the modern earth sciences, who advanced the scientific understanding of our planet and the processes by which it is constantly re-shaped.

The Fossil Flora of Great Britain

Employed early on in his career by Sir Joseph Banks, the botanist John Lindley (1799–1865) went on to conduct important research on the orchid family and also recommended that Kew Gardens should become a national botanical institution. This pioneering three-volume work of palaeobotany, first published between 1831 and 1837, catalogues almost 300 species of fossil plants from the Pleistocene to the Carboniferous period. The geologist and palaeontologist William Hutton (1797–1860), with whom Lindley collaborated, was responsible for collecting the fossil specimens from which the 230 plates were drawn. The first serious attempt at organising and interpreting the evidence of Britain's primeval plant life, this resource is notable also for its prefatory discussion of topics such as coal seams and prehistoric climate. Volume 2 opens with a preface on coal, followed by descriptions of some of the fossil plants found therein (plates 80-156).

Cambridge University Press has long been a pioneer in the reissuing of out-of-print titles from its own backlist, producing digital reprints of books that are still sought after by scholars and students but could not be reprinted economically using traditional technology. The Cambridge Library Collection extends this activity to a wider range of books which are still of importance to researchers and professionals, either for the source material they contain, or as landmarks in the history of their academic discipline.

Drawing from the world-renowned collections in the Cambridge University Library and other partner libraries, and guided by the advice of experts in each subject area, Cambridge University Press is using state-of-the-art scanning machines in its own Printing House to capture the content of each book selected for inclusion. The files are processed to give a consistently clear, crisp image, and the books finished to the high quality standard for which the Press is recognised around the world. The latest print-on-demand technology ensures that the books will remain available indefinitely, and that orders for single or multiple copies can quickly be supplied.

The Cambridge Library Collection brings back to life books of enduring scholarly value (including out-of-copyright works originally issued by other publishers) across a wide range of disciplines in the humanities and social sciences and in science and technology.

The Fossil Flora of Great Britain

*Or, Figures and Descriptions of the Vegetable
Remains Found in a Fossil State in this Country*

VOLUME 2

JOHN LINDLEY
WILLIAM HUTTON

CAMBRIDGE
UNIVERSITY PRESS

CAMBRIDGE
UNIVERSITY PRESS

University Printing House, Cambridge, CB2 8BS, United Kingdom

Published in the United States of America by Cambridge University Press, New York

Cambridge University Press is part of the University of Cambridge.
It furthers the University's mission by disseminating knowledge in the pursuit of
education, learning and research at the highest international levels of excellence.

www.cambridge.org
Information on this title: www.cambridge.org/9781108068550

© in this compilation Cambridge University Press 2014

This edition first published 1833–5
This digitally printed version 2014

ISBN 978-1-108-06855-0 Paperback

THE

FOSSIL FLORA

OF

GREAT BRITAIN;

OR,

FIGURES AND DESCRIPTIONS

OF THE

VEGETABLE REMAINS FOUND IN A FOSSIL STATE

IN THIS COUNTRY.

BY

JOHN LINDLEY, Ph. D. F.R.S. &c.

PROFESSOR OF BOTANY IN THE UNIVERSITY OF LONDON;

AND

WILLIAM HUTTON, F.G.S. &c.

───────────

" Avant de donner un libre cours à notre imagination, il est essentiel de rassembler un plus grand nombre de faits incontestables, dont les conséquences puissent se déduire d'elles-mêmes."—*Sternberg.*

───────────

VOLUME II.

───────────

LONDON:

JAMES RIDGWAY AND SONS, PICCADILLY.

1833—5.

BABINGTON, ———, Esq.

BRACKENRIDGE, GEORGE W., Esq., F.S.A., F.G.S., Broom-well House, Brislington, Bristol.

CONWAY, C. Esq., Pontycrum Works, near Newport.

COXWELL, G. S., Esq., Newcastle.

CUNINGHAME, Miss D'ARCY, of Lainshaw.

DICKSON, ROBERT, M.D., F.L.S., Licentiate of the College of Physicians.

DUNN, THOMAS, Esq., Newcastle.

EDGAR, THOMAS, Esq., 25, Dorset-place, Dorset-square.

EYTON, THOMAS, Esq., Eyton, Shropshire.

HAILSTONE, SAMUEL, Esq., F.L.S., Croft House, Bradford, Yorkshire.

HOBART, MISS, Lumley Park, Durham.

HURT, CHARLES, Jun., Esq., Wirksworth, Derbyshire.

IMAGE, Rev. T. Whipstead.

JOHNSTON, Dr. GEORGE, Berwick-on-Tweed.

KENT, WILLIAM, Esq., Bathwick Hill, Bath.

LAWSON, W., Esq., Brough Hall, Yorkshire.

LANGDON, AUGUSTUS, Esq., M.R.I., F.A.S., and Z.S.

LOONEY, FRANCIS, Esq., Oak-street, Manchester.

PRESTWICH, JOSEPH, Jun., Esq., F.G.S., Lawn, South Lambeth.

SAGE, Captain W., 48th Regiment, N. I. Bengal.

SAULL, W. DEVONSHIRE, Esq., F.G.S., 15, Aldersgate-street.

SIMPSON, Rev. J. P., M.A., Wakefield.

SWANWICK, Dr. Macclesfield.

TEALE, HENRY, Esq., Stornton Lodge, Leeds.

TYRCONNEL, the Earl of, Kiplin, near Catterick.

WALKER, Rev. ———, Ushaw College, Durham.

WETHERELL, N. T., Esq., F.G.S. M.R.C.S., &c., Highgate, Middlesex.

PREFACE

VOLUME II.

It was a part of the plan laid down when we
commenced this work, to take the opportunity
afforded by the appearance of each succeeding
volume, to state such general opinions as we might
be led to entertain on the subjects embraced;
accordingly, it is our intention at the present time
to detail some views we have been induced to take
of the circumstances under which the vegetable
fossils of the Carboniferous formation have been
deposited and mineralized, together with a general sketch of the rocks comprised in the term
" Coal Measures;" in the structure and composition of which, vegetable remains form so important a part, as to give an economical value to
them, far surpassing any other. In doing this,
we beg it may be held in view by our readers,
that our references will be made exclusively to

VOL. II.   b

the great Coal field of the North of England. We have several reasons for limiting ourselves, in the present article, to this district; the first is, it has been far more extensively worked, and its productions are, consequently, better known than any other. It has, also, furnished us with a very large portion of the materials we have hitherto made use of; and the residence of one of the Authors in the midst of it, has necessarily brought the circumstances attending it more particularly under our notice. There is a convenience, also, in thus limiting our references, as our observations cannot occupy a large space; besides which, we are convinced, that, in every essential circumstance, the history of one series of Coal measures is the history of every other of the same age.

It was our wish to have appended to this a Catalogue of all the vegetable fossils hitherto discovered in it; but, in attempting to form one, we have immersed ourselves in a labyrinth of difficulties, one half of its fossils having never been described; and, although we could easily ally a portion of these to known genera, yet the greater number of them would remain absolute riddles—waiting for some fortunate discovery by which they are to be connected with fossils already known, or proved to belong to others yet to be discovered.

The beds usually denominated the Coal measures, being the higher part of the Carboniferous

formation, occupy a large portion of the Counties
of Northumberland and Durham, reposing upon,
and being conformable to, the inferior members
of the series. They consist of irregularly alternating beds of sandstone, shale, or argillaceous
schist, and coal, whose aggregate thickness may
be estimated at 300 fathoms. This may not be
correct, but is, probably, near enough the truth
for our purpose.

With the exception of the coal itself, and a few
layers and nodules of clay-iron-stone, embedded
in some of the shales, the whole of these beds are
of mechanical origin, the shale being evidently
laminated clay, or mud, consolidated by pressure; and the sandstones abraded Quartz, Fel-
spar, and Mica, agglutinated by an argillaceous
or calcareous cement. From whence the im-
mense mass of travelled matter, of which these
sandstone and shale beds are composed, may
have come, it is somewhat difficult to conjecture.
The sandstones of the series below the Coal measures, denominated millstone grit, contain interspersed masses of water-worn quartz, of considerable size; and rarely amongst those of the Coal-
formation, a bed will be found, partaking of the
same characters; but the mass consists of minute
siliceous grains, which are not rounded, or but
partly so; from which it is fair to infer, that, whatever were its origin, the sand of which they are
composed was not brought from any great dis-

tance, or formed like the sands of our sea shore, by the slow action of attrition upon rocks previously consolidated, but that it had, probably, been produced by the ruin of crystalline rocks, so slightly coherent, as to have been unable to withstand the violent action of water, to which they had been exposed. The sandstones are all, more or less, micaceous, some of them containing that mineral in large quantity; where this is the case, and the plates are of considerable size, the stone is finely schistose. This is another proof that the materials forming the sandstone, had undergone little mechanical action previous to deposition, or the fragile mica would have disappeared.

In the series of beds, the coal itself forms, in bulk, a very inconsiderable portion of the whole. Forty seams are enumerated, but the greater part of them are too thin to be worked to profit.

The district has long been famous for producing coal of the finest quality, which has been extensively worked, and, up to the present period, the largest mining speculations in the kingdom, and, probably, in the world, are carried on within it. This being the case, it has become a matter of great economical importance, to define, as nearly as possible, each separate bed in the series, and this has been done with great minuteness. It is the universal belief of those best practically acquainted with the subject, that even the

thinner beds of coal, when not cut off by the rise
of the strata to the surface, or by some fault, are
spread out over the whole area of the formation.
Whether this be the case or not with all the
seams, we shall not stop to enquire; but the two
beds known as the High and Low Main Seams,
from their not only being the thickest, but as
affording, in their whole mass, coal of fine quality,
have been worked for centuries, and are known
over a space, in the first instance, of more than
80, and in the second, of 200 miles square.

In studying the Carboniferous formation generally, with reference to the circumstances under
which its different members have been deposited,
nothing is more singular than the sudden change in
the nature of the beds composing it, and the clearly
defined line by which these beds are separated
from each other; this is most particularly striking
in the lower portion, where a thick stratum of
Carbonate of Lime will be seen to terminate
abruptly, and be immediately succeeded by a
bed of entirely mechanical origin, and of a composition so opposite, as to contain scarcely any
calcareous matter whatever. Nor is the difference
of the nature of the two beds more striking, than
the difference of their imbedded organic remains;
whilst those of the limestone are almost exclusively of marine animals, the sandstones very
rarely contain fossils at all; and these, when pre-

sent, are, in a majority of cases, terrestrial vegetables.

The Carboniferous formation presents, from the
lowest to the highest member, a series of the same
vegetable forms. In the sandstone beds, immediately succeeding the old red Conglomerate,
which occurs at the base of the formation, along
the line of the great Cross fell fault, Sigillaria,
Lepidodendron, Calamites, and Stigmaria, begin
to make their appearance; as we ascend, the vegetable remains increase, whilst those of marine
animals, which existed in the limestone and shale
in profusion, decrease, until we arrive at the
Coal formation proper, where marine remains
disappear, giving place to those of vegetables
alone.

In this part of the series, we have the remains
of plants in every bed; the sandstones contain
them, but, from the roughness of their mechanical composition, it is the larger and stronger
stems only which have left their forms impressed
upon rocks of this class. Coal itself very rarely
retains any outward marks of its vegetable origin,
but the shale bed, immediately over the coal,
(when that substance forms the covering, as it
usually does,) furnishes us with fossils in the
greatest abundance. These are exposed by the
operations of the miner, who, in removing the
coal, often brings to light vegetable forms of sin-

gular beauty and variety, which are almost invariably found parallel to the laminæ of the stone,
and pressed flat, their outward form being retained on the shale as it was taken by the soft
mud which sealed them up, their substance being
converted into coal. Very large stems are often
found standing across the strata, and penetrating
through several different beds.

The vegetable origin of coal is now universally
conceded ; and it is almost as universally believed,
that the plants, of the remains of which it is composed, were swept by torrents from some neighbouring high and dry land, into lakes and estuaries,
where, becoming saturated with moisture, and
loaded with sand and mud, they sank to the bottom, and there reposed upon previously deposited
beds of sand and mud ; another vegetable mass
being in turn washed off, and buried by successive deposits of these substances, to be followed,
in due time, by another, and another.

Associated with the seams of coal, and in the
beds immediately surrounding them, stems of Sigillaria, of a large size, are frequently found
standing erect, with their roots proceeding from
them on all sides, (see vol. 1. plate 54.) We are
aware that the evidence of plants in this position
having grown on the spots where we now find
their remains, is not complete if taken alone, as
it has been argued they have been floated from a
distance, and left standing in an upright position

by the force of gravity, as is known occasionally to be the case during floods, where trees are removed along with the soil in which they grew; and this seems to have been certainly the case with the upright stems in the sandstone of the French mine of St. Etienne, where the different levels of their roots prove, as M. Constant Prevost has already remarked (*Dict. des Sc.* art. *Terrain,*) that they could not have grown where they now stand; but in the Lias Cliffs near Whitby, where the fragile stems of Equisetum columnare occur perpendicularly, they cannot have been so placed by force of gravity; and if evidence the most conclusive be required of the fact of vegetables having sometimes been overwhelmed on the spots where they grew during the deposition of the strata, it is furnished by the Fossil Forest of what is called the " Dirt bed," immediately over the fine building stone of the Island of Portland; and sub-marine forests of the present day supply us with the same fact, connected with a different order of things.

The fossils of the Coal measures occur often in groups; thus in the roof of the coal in Felling Colliery, the remains of Pecopteris heterophylla, (see vol. 1. plate 38,) were, a few years ago, most abundant; they occurred alone, almost unmixed with any other, over a considerable space, but, beyond that, have been rarely found, so that they are now comparatively scarce. Could such grouping have taken place if the individuals had been swept from a distance?

In plate 31, vol. 1, we figured a nearly perfect specimen of Stigmaria Ficoides, which was found, with two others, almost as perfect, in the shale forming the covering of the coal, in the Bensham seam, Jarrow Colliery, at the depth of about 200 fathoms from the surface; since that period, 14 others have occurred, all in the same bed, and within a space of about 600 yards square.*

Two of the specimens above alluded to, have been recently removed from the mine; one is the impression of the under side of the plant, shewing the central concavity, and 15 arms proceeding from it, four of which are distinctly branched; they are all truncated, the longest being four feet and a half.

The other specimen, of which the following is a sketch—

is of much smaller dimensions ; and, in this case, fortunately, the fossil has detached itself from the

* That a proper idea may be formed of the abundance in which the remains of Stigmaria occur in this bed, it should be stated, that those alluded to above, have all been brought to

roof, thus affording an opportunity of examining the upper surface of the central portion, which none of the before cited instances did. This exhibits the same wrinkled appearance, with indistinct circular spots, as the under side described vol. 1, page 104; it has nine arms, five of which sub-divide into two branches, at about 18 inches from the centre of the fossil, and one at three feet; in this, as in the other instance, they are all broken off short. This fossil, as before observed, occurred in the bed of shale immediately over the coal, towards which all the branches slanted. Two of these, which were longer than the others, were seen to reach the coal, where they were lost in the mass; whether the others had done so or not, could not be ascertained.

It would be out of place here, to recapitulate what has been already said of the form and nature of this strange fossil; but we must be allowed to observe, that the opportunities of further examination afforded by these several specimens, have proved that the centre was a continuous homogeneous cup, or dome, and not the remains of the

light in a short period, by the working of the mine; and that only in the roof of the passages, as from the mode of operation rendered necessary by the nature of the bed above the coal, at the first working, two thirds of that substance is left standing for its support; when this coal is afterwards removed, the roof will fall, so that it may never be possible to ascertain how many of these fossils now remain covered up.

arms squeezed into a single mass, as we formerly surmised it might be. We have, also, been furnished with the most convincing evidence of the leaves proceeding from the stem in all directions, thus :—

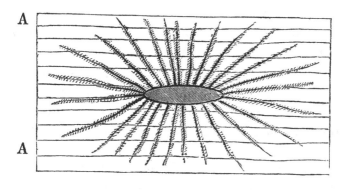

(A) Layers of Shale.

and, although we must still suppose the great length assigned to the leaves by that intelligent observer, Mr. Steinhauer, of 20 feet, to have originated in some error of observation, it gives us pleasure thus further to confirm the views originally taken by him, of this singular tribe of plants; we have, ourselves, seen the leaves well defined, three feet long.

Could it be possible for these plants, of a yielding fleshy substance, with numerous arms proceeding on all sides from a central dome, to be floated from the dry land, and buried in the mud

of an estuary, without being broken and squeezed
—the extent of the out-stretched arms, when perfect, having been at least 20 to 30 feet? If they
had been so floated, they must of necessity, in
sinking down upon the muddy surface, have become
flattened, and could not have presented the convex
form we now find them invariably in. The leaves,
also, which thickly surrounded the arms, could
not, under any circumstances, even supposing
them to have been hard woody spines, (which
they assuredly were not,) have taken the direction in which we now find them, proceeding from
the stem on all sides at right angles to its axis, and
penetrating the shale, even perpendicularly up and
down, to the extent of two or three feet, at least; had
the plants been floated, the leaves, on the contrary,
must of necessity have been pressed upon the
arms, surrounding which we should have found
their remains, in confused masses, and spread out
irregularly by their side, in the plane of the surface on which the plant had finally reposed; none
of this, however, takes place; but, on the contrary,
when the shale is split, so as to expose the surface of the fossil, the leaves are seen proceeding,
with the greatest regularity, each from its separate tubercle, those only being distinct in the
length and breadth, which, when in a growing
state, had been shot out in the plane which is
now the cleavage of the shale. (See plates 32
and 33, vol. 1.)

From all these circumstances, we are compelled
to conclude, that these Stigmariæ were not floated
from a distance, but that, on the contrary, they
grew on the spots where we now find their remains, in the soft mud, most likely, of still
and shallow water. It is worthy of observation, that the fossil remains of a Unio, (undescribed,) occur, in considerable abundance, associated with the Stigmariæ, but, in a shale, which
forms the covering of the high main coal in the
same colliery; and about 45 fathoms above the
Stigmaria bed, as we may very appropriately
designate it, there is, in one spot, a considerable
accumulation of this same fossil Unio : the coal
has been worked out under the layer of shells, in
all directions, and they are found to cover an area
of 5000 square feet. The shells are partly embedded in the coal itself, (which is spoiled by
them,) and partly in the shale above it; the bed
is about 18 inches thick ; the animals have, evidently, died at various ages; and the shells, of all
sizes, are, many of them, gaping open. As it is
impossible to conceive these, consisting of one
species only, to have been brought from a distance, and deposited here, we must conclude, that
this bed of shells, (and there are many more
known in other parts of the series,) marks what
had been, for some considerable period, as compared with the age of man, the uppermost surface
of the earth, upon which fresh, and, probably, still

water, had reposed, as in the before-cited case. Now, although it may be true, that the presence of organic remains in any stratum, be evidence sufficient of its having once been at the surface, yet the additional evidence in these cases, is so far valuable, as it proves that these beds remained uncovered for a period of considerable duration; long enough, indeed, for plants of a large size to flourish, and beds of muscles of considerable thickness to form, by the successive growth and decay of the animals.

What an amazing idea is thus forced upon us, of the length of the period which might elapse, during the deposition of the Coal measures alone, where the beds here referred to, are but two in hundreds, any one of which may have have been as long uncovered by its successors in the series; and what is the whole of the Coal formation, compared with the great mass of the secondary strata? —a single layer of stones in a stupendous edifice!

It has been already stated, that one of the seams of coal in the Northern Coal Field, is known over an area of 200 square miles; now, supposing this seam to have originated in the way generally believed, by a sweeping of vegetables from the land, could we, in any case, conceive such a mass floated down at one time, as to cover such a space? And if this bed be also spread over the formation where it has not yet been worked, we shall have to double or treble the space; if it had

been so produced, is it likely it would have presented, throughout the whole of this extent, an
absolute continuity, and an even thickness—this
thickness being, at the same time, so inconsiderable, as rarely to exceed six feet? Should we not
rather have expected to find the vegetable matter
unequally spread, and irregularly accumulated?

Again—if this seam of coal had originated in
the violent action of a current of water, sweeping
vegetables from the spots where they grew, would
not some of the soil and detritus in which they
vegetated, or the loosely aggregated matter which
then, at least periodically, existed in abundance,
be washed down and mixed with them? There
is no evidence of violent action whatever in the
beds of the Coal measures; there is not any thing
approaching a conglomerate, the grains of sand
comprising the sandstone being the largest trans-
ported fragments visible. It is one remarkable
character of the seams of rich coal, that, from the
floor to the roof, (to use the miners expressive
terms,) they contain no foreign admixture whatever. Occasionally, thin layers of sandstone, or
shale, occur, by which the seam is partially divided into two or more parts, indicating a slight
partial effusion of stony matter over the surface
of the vegetable mass, whilst it was yet forming;
but this is the exception to the rule; and only one
instance, that we are aware of, has ever occurred,
of a rolled fragment of stone being found in the

coal, and that was a pebble of water worn grey quartz, in Backworth Colliery, near Newcastle; we may be tolerably certain that such a circumstance is not common, as the high character of the Newcastle coal arises, in part, from the total absence of foreign matter.

Other arguments, to prove that the plants which formed coal were either not drifted at all, or at least not from any great distance, may be found not only in the perfect state of the leaves of many Ferns, but in the sharp angles of the stems of plants which there is every reason to believe must have been of a very succulent nature, such for example as Favularia tessellata, tt. 73, 74, and 75 of this work; and many of the Sigillarias, some of which occur with their surface marked with lines and streaks so delicate, that a day's drifting would have injured them. Again, at t. 76, we have figured a cluster of the fruits called Cardiocarpon acutum; had these been drifted, one would think they must have been dispersed, instead of being collected into one spot, just as if they had fallen there from the plant that bore them.

That the fossils which we find irregularly interspersed in the sandstones, or shales, of this formation, may have, in some instances, originated from drifted vegetables, there is, perhaps, reason to believe; thus it may have been with Dicotyledonous trees, fragments only of whose stems have been traced 70 feet long, without either extremity being seen ; these we are sure must have grown upon a dry surface, and that surface have been unchanged for many years. And, in fact, they are found in just the state in which we should expect to find drifted stems, their limbs shattered, their bark beaten and rotted off, and their wood in a high state of decay. But that any considerable part of the plants which formed the beds of coal were drifted at all, appears, from the foregoing remarks, to be highly improbable; that they should have been brought by equatorial currents from the regions of the tropics, is perfectly chimerical.

When such a mass of vegetable matter as is now periodically brought down by the Mississippi, is deposited upon mud, or sand, of which the bottom of some of its branches, or bays, may consist, and is there covered by another bed of sand, or mud ; is it likely, that, if, at any future period, the Carbonaceous deposit should be removed, the surface of the beds, either above or below it, would be even and flat ? Would it not rather be found, that the interstices and inequalities which there must be betwixt the trunks of the trees, had been filled up by the matter which covered the mass, and that some of the stronger stems, having settled unequally, had stood out, penetrating the surrounding soft strata, either above, or below ? Something of this kind, under similar circumstances, must, at all times, have been the case ; yet, nothing like an indication of it attends our coal beds, for, not only are they, as before observed, free from the admixture of matter foreign to the formation, but the surfaces by which the coal is separated from the beds above and below it, are as even and well defined, as those of the limestones in the lower part of the series.

From the circumstances already related, we are compelled to the conclusion, that the beds of coal chiefly originated in vegetable matter which lived, died, and was decomposed, upon the spots where we now find it. The analogy of Peat, at the present day, naturally suggests itself; and, according to this view of the subject, we must consider each of our coal beds as having originated in an extended surface of marshy land, covered with a rank luxuriant vegetation. Should the length of time required for such an accumulation of vegetable matter suggest itself as a difficulty, it may be in part got over, when we bear in mind the fact of the enormous size of the individual plants, and that all those having any living analogues, sufficiently attest a much more rapid growth, consequent upon a heated humid atmosphere, than, at present, is any where known to take place. The difference is, probably, not greater betwixt the stunted growth of an Iceland vegetation of the present day, and the rank luxuriance of a tropical swamp, than between even the latter and the vegetation of the Carboniferous period.

The remains of Stigmaria are so abundant throughout the whole of the Carboniferous formation, that it is impossible to travel far along any road, without its form being detected by the practised eye. In some of the best and most closely observed instances of its mode of occurrence in the bed before described, the arms could be traced from the central dome, slanting downwards into the coal, where all trace of them was completely lost. Coal, which rarely bears any outward vegetable form, presents that of Stigmaria oftener than any other, and it is certainly one of the most abundant fossils of the whole formation; from which facts, we should appear to be fully warranted in considering, that the growth of plants of this class was one of the great means made use of by the Almighty Architect of the globe, in absorbing and rendering solid that excess of Carbon, which, it is believed, must, at the period of the formation of the Coal-measures, have existed in the atmosphere; thus rendering it fit for the support of animal life, and, at last, a proper habitation for man. We cannot contemplate this storing up such a mass of combustible matter, and the iron which always accompanies it in the depths of the earth, at a remote epoch, for the consumption and enjoyment of creatures, afterwards to exist on its surface, without being struck with the benevolence and wisdom manifest in the design.

Whilst contemplating a bed of coal as the product of vegetation swept from a higher level of dry land, the question is ever recurring—where was the land?—a question which, as far as we

know it, is impossible to answer, and which
might be considered alone sufficient to shake the
theory of the Coal-plants having been drifted from
neighbouring hills. We are well aware that this is
but one of a thousand questions in Geology more
easy to propound than to solve ; but, surely, there
ought to be some indication of those rocks, of
anterior formation, on which this mass of vegetation grew ; the surface that could supply so
much, could be of no inconsiderable extent.
That the plants had not been brought from a great
distance, is proved, by the perfect state of preservation of the most delicate filmy leaves. The
only rocks of the older formation, near to the
great Northern Coal Field, are the Cumberland
group, and the Cheviots ; but it is certain that
the former were protruded at a period long subsequent to the formation of the Coal measures ;
and, although there is in the case of the Cheviots
a want of evidence to carry us so far up in the
great series, yet we are sure that they rose, after
the deposition and consolidation of the older members, at least, of the Carboniferous formation.
The beds below the Coal measures, do now rise,
at their western edge, to a height somewhat
mountainous ; but here, again, we have proof
of a rising, long posterior to the formation of the
coal ; and they are, besides, a part of the series
we are considering, and are characterized by the
presence of the same class of vegetable fossils as
have, doubtless, formed coal.

There are three principal varieties of Bituminous Coal, each of which occur in the Northern Coal Field ;—viz. fine caking Coal, which is a crystalline compound, breaking into rhomboidal fragments ; Cannel, called, also, Splint, and Parrot Coal, which is compact and tough, breaking with a conchoidal fracture; and Slate Coal, which is a mixture of the two other varieties, in thin horizontal layers.

The finest caking coal, of which the Newcastle Coal Field principally consists, being, as before stated, a crystalline compound, its constituents must have been in a state of solution. Cannel, or Parrot Coal, often bears the impression of plants, as does the third variety; but it is possible to prepare slices of all of them so thin as to be transparent, which, upon examination by the microscope, show the tissue of the original vegetables very clearly ; Cannel Coal seems to retain it throughout the whole mass, whilst it exists in fine coal in small patches only, which appear, as it were, mechanically entangled.

By the microscopic examination of coal, a singular arrangement becomes visible ; a number of elongated tubular passages are found, filled with a beautiful wine-yellow coloured resinous matter, which is the most volatile part of the solid coal, being what is first driven off when coal is exposed to heat. Each variety of coal exhibits this structure in a greater or less degree, but fine coal the least, as, in it, the vegetable elements appear to form an almost perfect union. When the different varieties of coal occur together in the same seam, or bed, as they frequently do, they are not indiscriminately mixed, but have a well defined line of separation between them. In Wylam Colliery, near Newcastle, the principal bed of coal is, at its lower part, a fine splint, approaching Cannel, the middle and main part is Crystalline coal, and the upper part of the seam is a mixture of the other two, in alternate layers, thus presenting, in one seam, all the three varieties of the Newcastle district. But it is not the seams of coal only which exhibit these abrupt changes of nature, as small specimens may be gathered at the mouth of every mine, which, within the compass of an inch, will, upon their perpendicular faces, show alternate layers of fine crystalline coal, and coal destitute of crystalline structure. It is certain each bed of coal, and more particularly each separate layer in that bed, must have been placed in precisely similar circumstances since the deposition of the vegetable matter of which it is composed; and we cannot suppose that matter to have obtained any of its elements after it was buried in the earth, but rather that the difference between the several varieties of coal and recent vegetables, as shewn by analysis, must have arisen from the play of affinities which has taken place in the mass when reduced to such a state as to allow of motion amongst the particles, (the result of the most complete solution of the fibre being the finest coal, whilst in the indifferent varieties this motion appears to have been obstructed by the tissue,) from which it seems naturally to follow that the several varieties of coal arise from some difference existing, previous to deposition, and that difference is most likely to have been, originally, in the nature of the plants, of whose remains the coal beds consist. If we are right in this conclusion, we are thus furnished with an additional argument against the common opinion of the origin of coal ; if the vegetables had been washed from a distance, is it likely that the different kinds would have separated so completely, as to have produced the several varieties of coal, so distinct from each other ? often in layers, far too thick and continuous for us to suppose them to have originated, but from a multitude of plants of the same kind. However this may have been, we have little doubt of being able to pronounce, with tolerable accuracy, as the knowledge of the subject extends, what the plants were, the remains of which are of such incalculable value to us in the form of coal.

It was at one time believed, that the remains of Dicotyledonous woods did not exist in the Carboniferous formation ; but subsequent observation, aided by the power of the microscope, which has been applied with so much perseverance and effect, by our esteemed friend and fellow labourer, Mr. Witham, has enabled us to detect them in almost every quarry. Nevertheless, the great bulk of the vegetables, of what may emphatically be called the Carboniferous period, undoubtedly have been of the genera Sigillaria, Lepidodendron, Calamites, Sigillaria, and Ferns. The more woody plants, on the contrary, after being buried, were able to resist decay, until their fine tissue was completely filled up and sustained, by the gradual infiltration of mineral matter.

It is in consequence of the almost universal change into coal, which has taken place in plants of this period, that their internal organization is so obscure; but, fortunately for our science, individuals are sometimes found uncompressed, and retaining the form of their internal organization in considerable perfection.

Mr. Witham has thus, already, been able to detect the structure of a Lepidodendron, which was fortunately found by the Rev. C. G. V. Harcourt, and upon which we shall have to make some observations in the present volume. To this part of the subject we should wish to direct the attention of our friends, more particularly such as may be resident in those Carboniferous districts where Calcareous Spar, and Sulphuret and Carbonate of Iron, abound; it is only where mineralizing matter has been held in chemical solution in abundance, that we can expect to find the delicate and evanescent textures of the coal fossils preserved. By careful examination in such situations, and the aid of the microscope, the secret of their real nature will be revealed.

BOTHRODENDRON PUNCTATUM.

(corticated.)

From the roof of the High Main Coal-seam, at Jarrow Colliery.

This is the remains of some large plant, of which the scarred stems and the bodies that belong to the scars alone are left.

Upon the surface of the stem are discoverable a considerable number of minute dots, arranged in a quincuncial manner, something less than half an inch apart : and it is probable that those may be the scars of leaves; but at present there is nothing to prove that they were so.

At intervals of ten or eleven inches, the stem is marked with deep circular concavities, four or five inches across, at the bottom of each of which is a distinct fracture, indicating that something has been broken out; while the sides of the concavities have concentric marks, as if from the pressure upon them of rounded scales.

Fragments, of which we possess one, have been taken out of the cavities, and shew that they are the points of attachment of very large cones, consisting, as far as can be made out from what is left, of rounded polished scales, three-tenths of an inch thick, attached to a central axis, and fitting accurately to each other. Upon the whole, they have so completely the appearance of the base of such a strobilus as that of *Pinus Lambertiana,* that we cannot doubt that the plant belonged to the natural order *Coniferæ.*

In recent plants, however, we have nothing at all like this in the manner in which the cones appear: for it seems as if they grew from the old trunk ; unless, indeed, we are to suppose, of which there is no proof, that the plant knew no seasons, but grew with such rapidity that its branches had acquired, by the second year, a diameter of seven or eight inches.

Of all the anomalous forms that the Coal measures have afforded traces, this is, perhaps, the most remarkable, and the best made out as to its external structure.

BOTHRODENDRON PUNCTATUM.

(decorticated?)

———

From Percy Main Colliery.

This is, in size, and all other characters, so similar to the last, that we can discover little difference between them, except in the absence, in this specimen, of the quincuncial dots, found on the surface of the other. We presume this to be an accidental circumstance, and that the specimen in question has lost its external surface. The scars are not more than six inches apart; but this cannot be taken as a distinctive mark, unsupported by other peculiarities.

ANTHOLITHES PITCAIRNIÆ.

From the shale associated with the Low Main Coal, at Felling Colliery.

Perhaps it would be scarcely worth publishing such fragments as those now represented, if it were not for the sake of adding a new proof to those already known, of the existence of an extremely diversified Flora, and of many highly organized plants, at the period of the Old Coal-formation.

This is, beyond all doubt, the remains of the inflorescence of some plant; but it would puzzle the most ingenious speculator to find a single character in the fossil, upon which a positive opinion as to its original nature can be formed. It seems as if it had been half decayed before it was imbedded, and its parts of fructification have so blended together, that it is in vain to attempt even to describe them; all that can be said is, that

A 3

there is a tolerably distinct appearance of a calyx, which seems to have enclosed petals much longer than itself; this, taken together with the probability that it owes its preservation to its having been originally of a hard and indestructible texture, has induced us to name it as if it had been allied to some of the recent tribe of Bromelias, to which, especially the genus Pitcairnia, it has as much resemblance as to any thing else.

NEUROPTERIS UNDULATA.

———

From the upper sandstone and shale of the Oolitic rocks, at Gristhorp Bay, near Scarborough, where it was discovered by Mr. W. Williamson, jun., to whom we are indebted for the figure, together with a specimen, and the following memorandum.

" This plant appears to have grown to a considerable length ; as, in the specimen from which the accompanying drawing was taken, there is little or no variation in the thickness of the petiole, through a space of eight inches. The latter has a deep furrow running down the centre."

From this circumstance, it is evident that the specimen is preserved with its upper surface only exposed to view ; a circumstance which is so common, as to lead to the suspicion, that *the true cause of the general absence of remains of fructification in fossil Ferns, is the greater adhesion of their lower fructifying surface to the matter in which they*

A 4

are imbedded, than of their upper, which is generally smoother, and has less means of sticking to the matrix.

" Part of the pinnæ are often met with, but generally in pieces not more than three or four inches long. In the pinnæ, the rachis has a small indistinct line, or ridge, on each side, to which the central vein of each leaflet appears to be attached." It is, therefore, probable that the rachis was winged. " Next the petiole, the leaflets are smaller and rounder than those at a greater distance, which gradually elongate, and take an undulated form ; but they frequently vary very much in shape."

This is very nearly allied to N. *Dufresnoyi*, var. β, found in the slate quarries of Lodêves, in the Department of the Herault, which are referred to the new red sandstone formation. Adolphe Brongniart, indeed, considers that species to be simply pinnated ; but unless he had better specimens for examination than those he has figured, one does not see why it should not have been of as compounded a structure as this.

PECOPTERIS REPANDA.

From Jarrow Coal Mine.

We know no species of Pecopteris with which this can be confounded ; its very blunt leaflets, which are almost cordate at the base, and its undulated outline, together with the distance at which its veins are placed from each other, are all peculiar to itself.

We have not, at present, met with it in any other situation than that above mentioned.

HALONIA? TORTUOSA.

———

In sandstone, in a quarry near South Shields,
from a specimen furnished by Isaac Cookson, Esq.

Whatever this may have been, it is evidently
very distinct from any thing hitherto described.
Probably, the present specimen has been jammed
and distorted so much, as to have lost, in a great
degree, its original character, but enough remains
to convey some idea of its external structure.

It seems to have been a plant of small dimensions, the surface of whose stem was completely
covered with little processes, which, in falling
away, left minute quincuncial ill-defined spots,
that rapidly became separated and obliterated,
as the stem advanced in age. Among these spots,
at intervals of three-fourths of an inch every way,
were arranged little projections, the apex of which
was terminated by some appendage now lost.
The ramification seems to have been dichotomous,
but this is extremely uncertain.

The principal questions to answer, are, first—
What were the processes? and, second—what
were the projections? If the processes were
leaves, as appears probable, then the projections
will have been either the bases of old, or the points
of rudimentary branches; and in that case the
affinity of the fossil will be nearest with *Halonia.*
(See the next Article.) But if we suppose the
processes to have been analogous to the ramenta
of Ferns, then the projections may be considered
of the same nature as those we find in *Stigmaria,*
where they are plainly the bases of leaves. A
great objection to this view is, that the arrangement of the spots left by the processes is too regular for ramenta.

The only branch that is seen in the specimen,
will not enable a Botanist to say whether the
mode of ramification was dichotomous, or alternate. If the projections are the bases of leaves,
it may have been dichotomous; but if they are
rudimentary branches, it must have been alternate.

Under these circumstances, we are forced to
leave the specimen in a state of uncertainty,
which is unfortunately but too common in this
science.

HALONIA GRACILIS.

————

From the Coal measures of Low Moor, in York-shire.

At first sight one would be disposed to consider this a *Lepidodendron*, to which its rhomboidal scars give it a strong resemblance. But if we consider Lepidodendron as an extinct form of *Lycopodiaceæ*, we must limit it to those fossils in which the mode of branching was dichotomous, for no other kind of ramification is met with in recent *Lycopodiaceæ*.

Here, however, it is plain, from the numerous scars of branches, that they were arranged in an alternate manner round a common elongating axis, after the plan that now obtains in the Spruce Fir. In fact, if we compare this with a vigorous branch of a Spruce Fir, one year old, we shall find the resemblance very striking, even in the scars of the leaves.

For this reason, and for the sake of rendering our notions of the extinct Flora as definite as we can, the genus *Halonia* is proposed to comprehend all those fossils, in which, to the surface of *Lepidodendron*, is added the mode of branching of certain *Coniferæ*, and which it is, therefore, to be inferred, were of a nature analogous to the latter.

CARPOLITHES ALATA.

From Jarrow Colliery.

It seems hopeless to determine the affinities of fossil fruits, unless they can be procured attached to the branches that bore them : for it is, in general, impossible, from external inspection only, to tell the relationship even of recent fruits.

For this reason we will not occupy time in profitless speculation upon the fossil plants to which these seeds have belonged, but confine ourselves to one point only.

It has been suggested that they are the remains of the seeds of some of the gigantic *Coniferæ* that flourished in the primæval forests, from the destruction of which coal has been produced ; and one would certainlyexpect to meet with both their cones and seeds, wherever the branches, which are the most perishable part, have been preserved. But up to the present day, we believe, that no one has found any trace of such parts, except in the curious case of *Bothrodendron*, (see page 1) ; unless some of the *Lepidostrobi* are considered Coniferous.

We cannot say that the fruit now represented is likely to have belonged to any of the extinct Pines; at the same time one would be hardly justified in absolutely denying it. Fig. 1, represents the fossil in a nearly complete state, with the outer shell unbroken ; but there is nothing to shew whether the shell was pericarpial or seminal. At Fig. 2, it is partially broken, so as to shew an internal cavity in which a round body is visible, which may have been either a seed or a nucleus ; from the twisted appearance of the surface of a part of this specimen, we may conclude that the shell was of spongy texture. Fig. 3, represents the nature of the internal cavity in a still clearer manner ; and it is evident that from the thicker end, where the seed lies to the narrow end, there was either a passage, or a vascular communication. In the former case, it might have been Coniferous ; in the latter, it must have been of a totally different kind, and the specimen must be considered inverted.

In point of size, the only recent Coniferous seed that can be compared to this, is that of *Arau-*

caria, one of which, from *A. Dombeyi*, is represented at Fig. 4, for the purpose of shewing how
little resemblance there is between even this and
the fossils in question.

ARAUCARIA PEREGRINA.

———

Communicated with the following fossil, from the Blue Lias of Lyme, in Dorsetshire, by the Misses Philpot.

The specimen, which has been carefully cleaned from the lias when soft, is one of the most perfect that we have ever seen; every thing, even the surface of the leaves, having been completely preserved. Unfortunately, the accompanying figure is not so good as could be wished; but we trust that any defects in it will be supplied by the following description of the specimen.

It consists of a branch upwards of a foot long, from the sides of which proceed four or five laterals, spreading widely from the main stem, and slightly curved. Both these, and the principal stem, are closely covered by thick, ovate, blunt leaves, which seem to have had a very broad edge, and a rhomboidal figure, and which

over-lap each other nearly half their length ; when
fresh, the leaves were probably even on the surface, but now they are a good deal shrivelled,
as if they had been half decayed when imbedded,
and their midrib projects till it reaches the apex,
which is slightly curved inwards; the whole surface is marked by minute impressed dots, like the
elytra of a colopterous insect.

Although the specimen is in good preservation,
and of large size, yet no trace of fructification is
discoverable on it.

The imbricated leaves remind one of the surface of *Lepidodendron;* but their thickness and
bluntness, and the want of all tendency to a dichotomous ramification, render it improbable that
the specimen was much related to that genus.

It is no doubt to Coniferæ that it is to be referred ; and in fact it is so similar to the adult
specimens of *Araucaria excelsa,* the Norfolk Island
Pine, that at first we fancied we should have a
case of identity between a fossil and a recent
plant.

But upon comparing the two plants carefully,
it turns out that the leaves of the fossil are so
much larger and blunter than those of the recent
species, as to leave no doubt of their having been
specifically distinct.   At the same time, the comparison confirms their great similarity, and estab-
lishes the important fact, that at the period of
the deposit of the lias, the vegetation was similar

to that of the southern hemisphere, not alone in the single fact of the presence of Cycadeæ, but that the Pines were also of the nature of species now found only to the south of the Equator. Of the four recent species of Araucaria at present known, one is found on the east coast of New Holland, another in Norfolk Island, a third in Brazil, and the fourth on the south eastern Alps of the American Continent.

STROBILITES ELONGATA.

From the Blue Lias of Lyme, in Dorsetshire ; communicated from the Museum of the Misses Philpot.

This remarkable fossil has occurred in a rounded mass of the Lias, the fracture of which has discovered it. It was evidently a cone formed of broad imbricated scales, which were longer about the middle of the cone than either at the base or apex. The scales in front of the specimen having been imbedded in the lias, are broken off, and nothing remains of them but their fractured bases ; but from the impressions of those at the side, it would seem that they had rather a lax arrangement, and were broadest at the point of attachment to the axis, that they tapered to the points, which were a little recurved, and that these points were abruptly truncated. This structure is sufficiently visible in some parts of the accompanying figure ; but it is much more perceptible in the fragment that corresponds with the part now represented ; from this fragment we are able to discover that the lower scales were not only shorter, but also thinner than the upper. No trace of the original surface remains·; but in its room, a thin stratum of cracked and broken carbonaceous matter overlies all the parts.

We presume there can be little doubt of this being a cone of some kind; and if so, it must have belonged either to some Coniferous genus, or to one of the Cycadeæ; for no other natural orders bear cones of such a kind.

To which of these it is to be referred, can scarcely be a matter of doubt. The great breadth of the scales at the point of their insertion into the axis is at variance with the structure of *Zamia*, to which alone, among Cycadeæ, the fossil can be compared; but it is in perfect accordance with that of Coniferæ, whether we contrast the specimen with the narrow cones of *Pinus Strobus*, and its allies, or with the broad ovate ones of such plants as *Araucaria* and *Cunninghamia*. It is, however, far from agreeing with any modern species, from all which its tapering but truncated scales distingush it essentially.

Is it possible that it can be the fruit of the plant last figured? This must of course be mere conjecture, there being no sort of evidence either for or against the supposition. It is nevertheless deserving notice, that supposing that plant to have been related to Araucaria, this fruit is of the same nature as it would in that case have been likely to have borne.

CYCLOPTERIS OBLIQUA.

Cyclopteris obliqua. *Ad. Brongn. Prodr. p.* 52. *Hist. des Végétaux Fossiles.* 1. 220. *t.* 61. *f.* 3.

Cyclopteris auriculata. *Id. Prodr. p.* 168.

Specimens of this extremely well marked fossil are not of very uncommon occurrence, but they do not seem to have been met with out of England. M. Adolphe Brongniart figured it from Yorkshire specimens, given him by Mr. Greenough; those now represented are from Jarrow Colliery; and we have received a drawing of a small specimen from Mr. Conway, found in the mines of Pont-newydd, near Newport, in Monmouthshire.

It appears to have varied a good deal in size, our fig. A being of the natural dimensions, B about a quarter less than the natural size, and Mr. Conway's much smaller than even A.

There is no living plant with which this can be identified, nor any fossil species for which it can be mistaken, the singular manner in which the base is hollowed out giving it almost the appearance of a human ear. It is not certain whether it was a simple leaf, or only a division of a compound leaf; but the want of any stalk to the base, in room of which there is the trace of what appears to have been a distinct disarticulation, inclines us to the belief that the latter is the more probable; and if so, it must have been, when alive, one of the most remarkable of its tribe, far exceeding in its dimensions any recent species.

The veins all radiate and dichotomize from the very base, and in no case appear to run together into a midrib; thus answering to the structure on which the genus Cyclopteris essentially depends, provided the leaves were simple. But if they were compound, it would rather belong to the genus Neuropteris. See tab. 91 A.

NEUROPTERIS INGENS.

We have received this species from several different localities. The specimen figured is from Jarrow Colliery, and we have several others in nodules of carbonate of iron from the Yorkshire Coal field. They vary in size from two inches and a quarter to nearly three inches in length, by from an inch and three quarters to two inches and a quarter in width.

Their texture seems to have been membranous, if we can judge from the very filmy and delicate state of their impressions. The outline was rather wavy, and the apex rounded; the base was apparently heart-shaped, and more or less oblique. The veins are almost those of Cyclopteris ; that is to say, they radiate from one common point, with little or no tendency to run into a midrib; but in some species they decidedly do coalesce ; and the great resemblance the leaflets bear to those of

Neuropteris auriculata, leaves scarcely any room to doubt their having belonged to a similar plant.

In fact, it is not easy to say in what respect N. *ingens* differs from the species just mentioned; but we are nevertheless persuaded that they must have been specifically distinct, for the leaflets of the present plant are at least twice, and frequently nearly three times as large as the largest of those of N. *auriculata.*

Is it not possible that *Cyclopteris obliqua* and *Neuropteris ingens* may both be leaflets of the same plant, the former coming from the base, and the latter from the sides of the divisions of the leaves? like the roundish, auriculated, and oblong leaflets of *Neuropteris auriculata.*

CYCLOPTERIS DILATATA.

––––––

From Felling Colliery.

This appears to have been of a very thin and delicate texture, and of considerable size; we possess one specimen, containing two-thirds of a leaf, which measures eight inches in breadth; it is probably on this account that it is never found perfect.

The outline of this species varies from nearly orbicular to oblong, with the principal diameter parallel with the base; it has an undulated surface, and its base is closed by two deep and equal lobes, which overlap each other. The veins radiate and dichotomize from their common point, without the slightest tendency to form a midrib.

At first sight it might be taken for *C. reniformis*; but that species does not seem to have been of so delicate a texture, was not much more than one third the size, and had not its base closed up by two overlapping lobes; on the contrary, its lobes were so short, as not to meet by a considerable distance.

TÆNIOPTERIS MAJOR.

Found in the shale of the Gristhorpe bed, in the Oolitic formation, near Scarborough, by Mr. William Williamson, Jun., to whom we are obliged for an excellent drawing, and for the following note.

" The specimen is about five inches long, and two broad ; the midrib is strong, and has a line upon its centre which gives it the appearance of having been once angular." (This line is no doubt the furrow that always exists upon the petioles of leaves, and thus shews the impression to be that of the upper surface.) " Running out perpendicularly from this midrib are numerous veins, which are twice or thrice forked, first near the middle, and again near the margin, in which character it differs from *T. vittata*. Some of the veins are even four times branched. The lower extremity of the leaf is destroyed."

To this we would only add, that while *Tæniopteris vittata* is hardly distinguishable in its fossil state from the Indian *Aspidium Wallichianum,* the species now represented may be almost identified with our British Harts-tongue Fern, *Scolopendrium officinarum,* which may be found in every old well, unless indeed the base of the fossil should prove, when discovered, to be much more different than its apex is.

As it would be a highly interesting discovery if the identity of the fossil and recent species could be established, we especially recommend a search after more complete specimens of this plant to our indefatigable friends at Scarborough.

LYCOPODITES WILLIAMSONIS.

Lycopodites Williamsonis. *Ad. Brongn. Prodr. p.* 83.
Lycopodites uncifolius. *Phillips' Yorkshire.*

Found very plentifully in the Oolitic formation of Scarborough. Mr. Phillips mentions it both in the upper and lower Sandstone and Shale. Our specimens are from Mr. Bean ; our drawing from Mr. William Williamson, Jun., with the following note.

" This appears to have been a creeping plant, like our Lycopodium clavatum. The stem is frequently branched, and concealed by the base of the leaves, which are sessile, and of an acute falciform shape. Up the centre of each leaf there is one, and sometimes two strongly marked ridges, which have evidently been edges of angles. The leaves are placed opposite each other, and have frequently smaller ones situated between them. The surface of the stem is covered with scales apparently the base of leaves, which have lost their points. The stems are terminated by a large oval head, or cone, which is covered with small hook-like processes, similar in form to the leaflets, but smaller. Where the bituminous substance is destroyed, there are strongly marked rhomboidal spaces, looking like scars. Fragments of this plant are very plentiful, but attached heads are rarely met with ; the one figured is from Gristhorpe Bay."

No modern species can be compared with this for size, especially that of the heads, which are very much the same as the Lepidostrobi of the coal measures, fossils which probably belonged to similar plants.

What we find more especially remarkable in this species is, that, notwithstanding its great size, it must have belonged to the most delicate division of the genus, as is proved by the stipulæ accompanying its leaves. The largest Lycopodia of the present day have leaves without stipulæ ; but in the days when the Oolitic rocks were deposited, things must have been ordered differently.

Mr. Williamson has drawn a specimen, in which the main stem terminates in a cone; it often happens that the lateral branches also bear cones, but in that case the former are so very short, that the latter are almost sessile.

94

PECOPTERIS NERVOSA.

Pecopteris nervosa. *Ad. Brongn. Hist. des Végétaux Fossiles,*
t. 94. excluding Sternberg's Synonym.

In Shale from the Bensham Coal Seam, in
Jarrow Colliery.

This is evidently the same plant as is figured
(but not yet described) by M. Adolphe Brongniart,
at t. 94 of his great work, under the name we
have adopted; but we cannot think the synonym
of Pecopteris bifurcata right, as that species has
evidently veins far more wide apart, and a very
different outline; it is, however, in all probability,
the same plant as appears at t. 95. f. 1 and 2 of
M. Brongniart. The letter-press that refers to
these plates not having yet appeared, we are un-
acquainted with the motives that has led to the
combination of plants apparently so very different.

The appearance of this species calls to mind several kinds of Asplenium, but we have not discovered any one with which it is of importance to compare it.

KNORRIA TAXINA.

From the roof of the High Main Seam, in Jarrow Colliery.

Surely this must be a portion of the branch of a Yew, or of some such plant. Let it only be compared with the one year old shoots of that tree, the leaves having been stripped off, and something very like identity will be found to exist; especially in the manner in which the leaves ran down upon the stem, and in the nature of the scars they left behind.

To illustrate this, we have introduced some figures of the Yew branch of different ages, which may also be taken as explanatory of other cases of similar structure.

B. represents a very young Yew branch, with its leaves broken off.

C. is the same, a little older, and with the leaves fallen off naturally.

D. is another portion of a branch, much older;
at *a* the bark is stripped off, so as to shew the
difference between the corticated and decorticated
surfaces.

Knorria is a genus of Count Sternberg's, not
noticed by Ad. Brongniart; for remarks upon
which, see *t.* 97.

CALAMITES ———.

(The Base of a Stem.)

From the roof of the Bensham Seam, at Jarrow Colliery.

To what species this singular fragment belongs, we are unable to determine.

We only figure it for the sake of indicating what the nature is of the fossils that appear in this state.

Collectors should never trust to specimens of such a kind as illustrative of strata, but should, in all cases, take the middle or upper ends of the stem, in which alone that evidence can be found which is necessary for the determination of the species of Calamites.

KNORRIA SELLONII.

Knorria Sellonii. *Sternb. Flore du Monde primitif. fasc. 4. pp.* xxxvii. *&* 50. *t.* 57

From Felling Colliery.

This plant, in a more perfect state, with the leaves, and cortical integument nearly complete, has been figured by Count Sternberg from the Frederick Gallery in the coal mines of Saarbruck ; the same author cites England, and the grauwacke of the neighbourhood of Magdebourg, as also pro ducing it.

In its more perfect state it presents a broad even surface, covered with cylindrical processes, which are not further apart than their own diame- ter, or a little more. In the state now represented, in which the bark and cylindrical processes have altered their appearance from the wasting of the

stem before consolidation, the place of the processes is occupied by flattened projections, broke at
their ends, and marked by a very shallow furrow,
which passes from the point downwards, losing
itself on the surface of the stem.

Such a specimen as this would not throw much
light upon the original structure of the plant; we
therefore transcribe Count Sternberg's account of
those which he had examined.

" I formerly," he observes, " described another
species of this genus, under the name of *Lepidolepis*,
being at that time of opinion that traces of the
attachment of scaly leaves could be distinguished
upon its impression. I have, nevertheless, since
satisfied myself not only by the examination of the
present subject (*Knorria Sellonii*), but also by
others of a similar kind, that in these cases it is
not mere scars that are preserved, but real cylindrical leaves, like those now commonly met with
in succulent plants. In this case, they are partially broken. If the point of insertion were visible,
this plant would resemble a Variolaria (i. e. *Sigillaria*). There was this, in particular, in these
plants, that they were rounded at the top, like certain species of Euphorbia and Melocactus, where
a tuft of hairs, or something of a similar kind, terminated the plant. This circumstance I have observed in a *Variolaria* (Sigillaria) from Saarbruck,
and on a plant of the present genus in the organic
remains of Steinhauer. There can, therefore, be

little doubt that these were really the representatives of succulent plants in the primæval world."

The other species to which Count Sternberg refers in the preceding paragraph, formerly named by him *Lepidolepis imbricata*, and now *Knorria imbricata*, he had procured from the grauwacke at Magdebourg, and from the coal mines of Orenburgh, on the borders of Asia. We believe we may also refer to that plant some remains found in the sandstone of the Ketley Coalfield, in Shropshire, for a specimen of which (numbered 12) we are indebted to Mr. Lloyd.

The genus Knorria is passed by unnoticed by M. Adolphe Brongniart; and even the species referred to it by Count Sternberg are uncited in the Prodromus; we are, therefore, ignorant to what other genus M. Brongniart considers them reducible.

It is with Lepidodendron and Stigmaria that they have the greatest apparent relation, as far as external characters go; but if the opinions just quoted are well founded, Knorria must have been extremely different from the former; while the latter would be distinguishable by the round projecting tubercles out of which the leaves arose. We would, therefore, preserve the genus Knorria, and provisionally refer to it not only the two species of Count Sternberg, but also all fossil plants, the leaves of which were in a densely arranged spiral manner, and have left, not depressed but projecting, scars. It is no doubt true, that, by such a character as this, plants may be combined originally of extremely different appearance; but we are forced to admit such characters in the present state of our science, from want of others of a more positive kind.

LEPIDODENDRON HARCOURTII.

Lepidodendron Harcourtii. *Witham in .Trans. of Nat. History Soc. of Newcastle upon Tyne, March,* 1832. *Id. Internal Structure of Fossil Vegetables, p.* 51. *tt.* 12, 13.

This interesting fossil occurred in the roof stone of a bed of coal worked at Hesley Heath, near Rothbury, in Northumberland: it is there found a few fathoms below a thick limestone, which is by some considered analogous to the great limestone of Alston Moor: whether this be the case or not, the position of the seam must be deep in the mountain limestone series. The fossils are found partly in the coal, and partly in the roof, which, in many cases, consists of a mass of encrinal remains and shells, such as Productæ, Melaniæ, &c., with the exterior converted into pyrites, in contact with the coal. The fossil is mineralized with clay iron stone and iron pyrites, having a coating of fine coal.

It was originally found by the Rev. C. G. V. Vernon Harcourt, Rector of Rothbury, to whose liberality we are indebted for the inspection of several specimens of the beautiful internal structure which Mr. Witham has, fortunately for science, discovered to exist in it. By means of these, of others communicated by Mr. Witham himself, and of a portion of a stem belonging to the Yorkshire Philosophical Society, for which we are obliged to Mr. Phillips, we have been enabled to prepare the following account of what is, beyond all doubt, the most remarkable discovery in the science of Fossil Botany.

The structure of this plant has already been so carefully described by Mr. Witham, firstly, in the Transactions of the Natural History Society of Newcastle upon Tyne, and, secondly, in his valuable observations upon the internal structure of Fossil Vegetables ; and the figures that accompany the description of this indefatigable geologist are so perfect, as to render it unnecessary for us, on the present occasion, to do more than select some of the more important parts of structure for representation, referring those who wish to consult more extensive figures to the publications just mentioned. There are also two or three points upon which we hope to be able to throw some additional light.

The stem seems to have been from an inch and a half to two inches in diameter, and of a cylindrical figure, producing forks occasionally. Its surface was marked with scars, arranged in a spiral manner, having the usual rhomboidal or oval figure of other Lepidodendra, but not sufficiently well preserved to shew precisely what their configuration was; they seem to have had a furrow running down their middle. Over the whole of these is now found a layer of carbonaceous matter, which is probably foreign to the stem itself, as it exhibits no trace of structure, and is apparently unconnected with the tissue which it will presently be seen that the stem still consists of.

When cut across and polished, the centre of the stem is frequently found converted to calcareous spar, which also has filled up irregularly a vast number of curved passages, proceeding upwards and outwards from the centre to the scars upon the surface. These curved passages give the stem, when sliced in a direction parallel to the surface, the singularly mottled appearance which is represented in Mr. Witham's Plate xii. f. 3 and 4. All the other part of the stem is hard and black, and distinctly organized, the calcareous spar chiefly indicating the parts where the tissue is obliterated.

When viewed with the microscope, the following appearances present themselves. Next the surface a horizontal section shews a dense layer of quadrangular meshes, very like those in Coniferæ, with irregular circles lying among them, also similar to the fistulæ of the same tribe of plants; this dense layer of meshes passes irregularly and insensibly into an extremely lax kind of cellular tissue, which extends from this point to the axis, constituting the principal mass of the stem. (See tab. 99. fig. 2. where a is the outside, b the inside, and c the irregular circles. A vertical section of this same part, shews that none of the above described meshes are the mouths of tubes, but that they are merely sections of cellular tissue, of which only that next the outside is elongated perceptibly in the direction of the axis. (See tab. 99. fig. 3, where a is the outside, and b the inside.)

The centre of the stem, or the axis, when viewed horizontally, is found to consist of a column of very lax cellular tissue, the innermost part of which is obliterated by calcareous spar; on the outside of this is placed a circle, consisting of much more compact cellular tissue, in which lie, at nearly equal distances, and next the outside, a considerable number of oval spaces, (Tab. 99. fig. 1. $a.$) composed of a fine net-work, bordered by a colourless ring, the structure of which is not determinable. A vertical section of this part shews that the fine net-work in the midst of the colourless ring is the mouths of vessels *having most distinctly a spiral structure.* The appearance of these vessels, when very highly magnified, is given at tab. 99. fig. 4. What the colourless ring was, is not discernible from the specimens ·we have examined, but in all probability it was the tube of woody fibre, which, in recent plants, usually accompanies and protects the bundles of spiral vessels. This part of the structure, Mr. Witham does not appear to have met with in his specimens.

On the outside of the vascular sheath just described, are occasionally to be seen little oval spaces, composed of net-work like that within the colourless rings, (see tab. 99. fig. 1. *b*.); they have been figured at his Tab. xiii. f. 4, *a*, and f. 2, *e*, by Mr. Witham, who considers them bundles formerly surrounding pith, (p. 53.) From their position, and from the irregular distance at which they are placed round the vascular sheath, they were, we think, more probably the mouths of the vessels, which it will be presently seen exist, in the curved passages already spoken of.

From the centre to the circumference, obliquely upwards and outwards, proceed a great multitude of these curved passages, which evidently correspond in number to the scars of the leaves, in which they also terminate. A section of one of these passages, made at right angles with its line of growth, exhibits two clusters of meshes placed one above the other, each surrounded and separated from the other by a fine and nearly obliterated net-work, which itself lies in the midst of the coarse cellular tissue of the stem. (See tab. 99. fig. 6. *a. b.*) An oblique section of a passage shews also that the clusters of meshes were the mouths of two bundles of spiral vessels, the upper bundle being much larger than the under, (see tab. 99. fig. 7. *a. b.*); this is confirmed by a longitudinal section of the same part, where the vascular structure becomes beautifully and distinctly manifest, (see tab. 99. fig. 8. *a. b.*); in this and the last case, the vessels are evidently surrounded by a sort of fibrous matter, which is probably woody fibre, the mouths of which produced the fine and nearly obliterated net-work of fig. 6. Sometimes all trace of this organization is destroyed, and the oblique sections of the passages are partially filled with unorganized carbonaceous matter, as at tab. 99. fig. 5.

We believe the whole of the foregoing description is essentially in accordance with Mr. Witham's observations, with the exception of the vascular sheath described as surrounding the central cellular column or pith; a point, however, to which great importance must be attached.

Such being the structure of this plant, we have thought it might not be entirely useless if we introduced into one of our plates an ideal view of its tissue restored to what may be presumed to have been its state when growing. This will be found at tab 98. fig. 2, where the figures and letters all correspond with the same figures and letters in tab. 99, so as to shew at once the evidence upon which the restoration has been made. Unfortunately, the vertical section next the bark is not represented quite as we could have wished, for the diameter of the elongated cells is greater in the vertical than in the horizontal section, and there are also some other points in which our engraver has not been so faithful as we could have desired. We trust, however, that the figure will answer the purpose for which it was intended.

The next point for consideration is, how far the discovery of the internal anatomy of this plant confirms the opinions previously entertained of the analogy of Lepidodendron to recent plants.

It has been generally admitted that this genus was related to Lycopodiaceæ; it has even been believed to be identical with the recent Lycopodium; and Mr. Witham considers that there is nothing in the structure of the present species that might tend to invalidate the opinion.

It is, however, no small gratification to ourselves to find, that all which we said upon the subject at page 19, &c. of our first volume, is completely confirmed ; and that it " is not exactly like either Coniferæ or Lycopodiaceæ, but that it occupies an intermediate station between those two orders," &c. vol. 1. p. 21.

It had a central pith, it had a vascular sheath surrounding that pith, and it had fistular passages in its cortical integument; thus far it was Coni-

ferous. But no trace can be found of glandular
woody fibre; it can scarcely be said to have had
any wood; and it is uncertain whether it had
bark; if it had, the bark must be considered to
extend from the external surface to the vascular
sheath; nor is there even in recent Coniferæ
such distinctly marked curved passages, connecting the leaves with the vascular sheath; curved
passages, no doubt, exist in Coniferæ, but they
form a very inconsiderable proportion of the vascular system.

Its vascular system was confined to the middle
of the stem, and to the curved passages emanating
from it; the stem consisted chiefly of lax cellular
tissue, which became more compact towards the
outside, and it had a very powerful communication between the bases of its leaves and the central
vascular system; thus far it was Lycopodiaceous.
But recent plants of the latter tribe have no fistular cavities in their cortical integument: a point of
great importance, because such cavities indicate
the presence of resinous or other secretions, which
are never found in Lycopodiaceæ; and, secondly,
the latter have no vascular sheath surrounding
pith, which is a sure sign of a dicotyledonous
structure, and quite at variance with the plan
upon which Lycopodiaceæ are organized. In *Lycopodium rigidum* the axis of the stem consists of a
bundle of five or six large spiral vessels, sur
rounded by four or five layers of smaller ones;

on the outside of this is a rather compact layer of cellular substance, which is connected by very lax cellules with the cortical integument, which is again more compact; the same structure exists in *Lycopodium cernuum*; and Mr. Witham represents a nearly similar arrangement of parts in *Lycopodium clavatum.* Not a trace of pith, or of the preparation for it, can be found.

We may, therefore, conclude that Lepidodendron was intermediate between Coniferæ and Lycopodiaceæ, constituting the type of a kind of structure now extinct. To Botanists, this discovery is of very high interest, as it proves that those systematists are right who contend for the possibility of certain chasms now existing between the gradations of organization, being caused by the extinction of genera, or even of whole orders; the existence of which was necessary to complete the harmony which it is believed originally existed in the structure of all parts of the Vegetable kingdom. By means of Lepidodendron, a better passage is established from Flowering to Flowerless Plants, than by either Equisetum or Cycas, or any other known genus.

SPHENOPTERIS CRENATA,

AND

SCHIZOPTERIS ADNASCENS

———

This very remarkable fossil was found in the shale of the Whitehaven Coalfield, from which it has been obligingly communicated to us by Mr. Williamson Peile.

It is evidently formed by the association of two distinct plants; one of which is a fern, around the stem of which another plant, possibly a fern also, has twisted itself. They are of totally different structure, and require to be described separately.

SPHENOPTERIS CRENATA.

By this name we would designate the principal fern in the accompanying plates. It was apparently a plant with a tripinnated leaf, the ultimate

segments of which had a narrow lanceolate tapering outline, and a regularly crenated or obtusely
lobed margin; these segments adhered to the
rachis by the whole of their base, and did not
exceed two and a half lines in length at that part;
towards the point they became gradually smaller
till they were reduced to a single lobe. Plate 100,
represents it half the natural size, and Plate 101,
its full size.

In the specimens we have examined, the veins
are totally destroyed, except a faint trace of a
midrib, which passes from the base to the apex of
each segment; of lateral veins no indication can
be found.

We have referred it to the genus Sphenopteris,
chiefly on account of its general resemblance to
S. Dubuissonis, from which it is distinguished by
its smaller size, and the entire crenatures, or lobes
of its segments.

SCHIZOPTERIS ADNASCENS.

To the obscure genus Schizopteris, we refer the
plant that is twisted round the stem of what we
have just described. Up to the present time, no
authentic figure has appeared of the genus which
M. Adolphe Brongniart has thus designated; but
we presume the " *Filicites crispus*" of Germar and

Kaulfuss, is one species; and, if so, this must be
another.

It may be conjectured to have been of the
nature of some of the Lygodia, or rather Hyme-
nophylla; and that the deeply lobed bodies,
of which the impressions are left, were the
leaves. They were palmated and divided into a
number of narrow segments, which sub-divided
into two or more commonly three lobes, which
were either entire, or forked, and always sharp
pointed. No trace of veins can be discovered,
unless the delicate striæ with which the whole
surface of the leaf is covered, be considered such.

In whatever way we look at this fossil, it cannot
but be considered important, as indicating a climate of tropical character. The only recent ferns
to which the Schizopteris can be compared, are
tropical, or nearly so ; but we have not, as far as
we know any modern instance of one fern twisting
round another, although it is possible to conceive
that such a thing might happen with such plants as
Lygodium. If it did happen, it is at least certain,
that the growth of the climbing plant must have
been as rapid as that of the species to which it is
supposed to have been similar; and that its vegetation must have been stimulated by a climate extremely different from that of Great Britain, at the
present day. For it must be remarked, that, in
this country, a few leaves closely collected round an

exceedingly short stem are all that one season is able to produce, while, in the case of the *Schizopteris*, not only must a considerable number of leaves, but also a great extent of stem, have been produced in that period.

102

PTEROPHYLLUM PECTEN.

Cycadites Pecten. *Phillips's Yorkshire.*

From the rich bed of fossil plants in the Oolitic formation of Gristhorpe Bay,* near Scarborough; for the communication of which, we are obliged to Mr. W. Williamson, jun.

It appears to have been a species of Cycadeous plant, as far as can be made out from the remains that have been discovered. We refer it to the genus Pterophyllum, because it appears to have more relation to that than to any other; but it is necessary that the technical character of that genus

* We are under obligation to our excellent friend and correspondent, Dr. Murray, of Scarborough, for many interesting specimens of the fossils found in the rich deposit of Gristhorpe Bay. His fine collection from that locality, has frequently been of essential service to us.

should not be made to depend upon the form of its pinnæ, but upon its veins being all of the same size, and the segments of the leaves attached to the midrib by their whole base.

We have nothing recent to compare it with. Mr. Williamson's account of it is as follows :—" The midrib of this elegant little plant is about one-eighth of an inch in width, tapering gradually, and is terminated by a small blunted segment. There are some traces of longitudinal striæ upon it, but so small, as to be nearly imperceptible. The segments are extremely regular, placed alternately, and thickly covered with very fine parallel veins; I think they are simple, but being very indistinct, I cannot be certain."

103

CTENIS FALCATA.

Cycadites sulcicaulis.   *Phillips's Yorkshire.*

From Gristhorpe Bay.

To Mr. Williamson we are again indebted for our knowledge of this curious plant, upon which' he makes the following remarks:—

" The stem of this plant is about one third of an inch in breadth, straight, of an equal width, and terminated by a lanceolate segment; its surface is covered with longitudinal striæ, from whence Mr. Phillips named it. The leaflets are numerous, linear, broadest at their base, and tapering to a narrow pointed apex. The veins run parallel with the edges, and are frequently forked, as seen in the magnified portion; at the junction of the leaflets to the midrib, the veins diverge in opposite directions, as will be observed

E 4

at *a a*. This plant is found, not unfrequently, at Gristhorpe Bay."

Mr. Phillips refers it to Cycadites, but to this the forked veins offer, what we fear, is a fatal objection. It is, however, difficult to say, to what else it can be better compared, unless to some ferns, such as *Acrostichum alcicorne*, in a fertile state. To this, however, there is an objection; for, while A. *alcicorne* evidently owes the peculiar arrangement of its veins to an extension of a leaf in which the usual forked structure exists, this fossil can scarcely be considered otherwise than as representing the general character of all the leaves of the plant.

It is not impossible that it may have belonged to some Palm ; but as there is no kind of evidence that this was so, we prefer placing it in a provisional genus, for which we venture to propose the name of *Ctenis*, in reference to its pectinated character. To this we would refer all leaves having the general character of Cycadeæ, but with the veins connected by forks, or transverse bars.

DICTYOPHYLLUM RUGOSUM.

Phyllites nervulosus. *Phillips's Yorkshire, t. VIII. f. 9.*

First described by Mr. Phillips, from the upper sandstone, shale, and coal, of the Oolitic formation, in Yorkshire. Our drawing was communicated by Mr. Williamson, jun., to whom we have so often had to express our obligations.

It was evidently a pinnatifid leaf, belonging to some exogenous plant; but to what recent species it may be analogous, it would be idle to inquire, so common are its form, and the arrangement of its veins. It might have belonged to a tropical or a European genus, to a tree or a herb, to a Sowthistle or a Scrophularia—in short, to plants of the most opposite qualities and structure.

If the genus Phyllites, in which Mr. Phillips has

placed it, be taken as the receptacle of all sorts of leaves, it will prove so heterogeneous an assemblage, as to cease to possess any precise character. Although it seems hopeless to determine the exact analogy of the greater part of the Monocotyledonous and Dicotyledonous plants of which the leaves alone can be found; yet important geological objects may be obtained by such a nomenclature of leaves as shall not violate natural affinities, and shall enable them to be accurately identified. We would, therefore, confine the term Phyllites to those Monocotyledonous leaves in which the principal veins converge at both the base and apex. For doubtful Dicotyledonous leaves of common reticulated structure, such as this, the name Dictyophyllum might be advantageously employed; and other names might be invented for leaves having remarkable peculiarities in the arrangement of their veins.

NEUROPTERIS ARGUTA.

From Gristhorpe Bay, communicated by Mr. W. Williamson, jun., with the following description :—

" The rachis is nearly smooth, broad at the base, and tapering gradually towards the apex. The pinnæ are oblong-lanceolate, pinnated, and tapering gradually from the base upwards, until they end in a very narrow point. The leaflets are oblong, and attached obliquely by a part of their base, with a slightly wavy margin. The central vein of the leaflets is very strong near the base, but disappears before reaching the apex ; the central veins are forked, curved, and set obliquely upon the central one. Towards the upper part of the leaf, the leaflets become much more acute, and the leaf itself is terminated by segments, like those of the lower pinnæ."

" At first sight, it would seem as if the fragments now represented, were parts of two diffe-

rent species ; but, in a specimen which I have
seen, both form part of the same impression,
proving them to be the two extremities of the
same half."

PECOPTERIS INSIGNIS.

———

Found at Gristhorpe Bay, in a nodule of iron-
stone, by Dr. Murray and Mr. Williamson, sen.,
of Scarborough.

It appears to have been a very beautiful species,
and was, probably, of a larger size than is usual in
the Oolitic formations. Mr. Williamson, jun., to
whom we are indebted for the drawing, describes
the main stalk as being a quarter of an inch wide,
and deeply furrowed in an irregular manner. The
leaflets are about an inch and a half in length, of
a narrow lanceolate figure, set on the rachis by
their whole base. The secondary veins are planted nearly perpendicularly upon the midrib, and
fork with great uniformity.

There is no published species to which this even
approaches.

PECOPTERIS SERRA

———

In shale, from the Whitehaven Coal-field, communicated by Mr. Williamson Peile. We have been favoured with fine specimens from the Natural History Society of Newcastle.

Those we have examined, have very much the appearance of some modern *Pteris*, and, probably, belonged to a plant not very different. All that remain are fragments of what seem to have been divisions of a tripinnate leaf of considerable size, the final segments of which had a long linear lanceolate figure, with about 20 or more lobes on each side. These lobes are at the bottom, of an ovate oblong form, attached by their whole base to the rachis, a little curved forwards, and very slightly wavy at the margin. Their veins are badly preserved ; but it would seem as if there had been a perfect midrib, upon which forked veins were planted almost perpendicularly.

But a small part of the extensive genus Pecopteris is yet published; from all the species of which
any figures have been given, this is widely different;
but we are uncertain whether it may not be
already named by M. Adolphe Brongniart, in his
Prodromus.

ASTEROPHYLLITES COMOSA.

From the shale of Jarrow Colliery.

This occurs in extremely indistinct impressions, of which nothing but the outline of the leaves remains; they were numerous and regularly verticillate; their figure was exceedingly narrow, and there is no perceptible trace of any kind of vein. The stem which bore them has, also, disappeared, leaving not a vestige even of its surface.

The genus Asterophyllites is so vague, that it will comprehend any fine-leaved verticillate plants, the bases of whose leaves do not run into an annular rim. For this reason we refer this fossil to it, although it is not improbable that it may be essentially different from those we have already described under the same generic title. It would be a bootless inquiry to attempt to discover a modern analogue; for so totally destitute of positive information are the remains, that five hundred plants might be named, to all which they would be extremely similar, and yet, perhaps, essentially distinct from all.

The three broader linear leaves which seem to rise from the base of the specimen, have nothing to do with the species, but are the remains of some *Poacites*, which have, evidently, been in contact with the Asterophyllites itself, at the time it was imbedded.

SPHENOPTERIS OBOVATA.

In shale, from the Newcastle Coal-field; drawn from a specimen presented for this work by the late T. Allan, Esq., Lauriston Castle, Edinburgh.

It occurs in small terminal fragments, which are arranged as if they were the lateral divisions of a tripinnate leaf. The rachis is, in all cases, nearly destroyed, nothing of it being left beyond a deeply sunken furrow. The final pinnæ have an oblong lanceolate figure, and are divided into about six obovate segments. No midrib can be found on these segments, nor any other kind of veins beyond a number of very fine parallel striæ which occasionally fork.

There is no species yet discovered with which this can possibly be confounded.

A FOSSIL AQUATIC ROOT

———

Myriophyllites gracilis. *Artis, Antediluvian Phytology, t.* 12.

———

A rare fossil, found in the low main of Felling Colliery, whence our specimen was procured; and, also, according to Mr. Artis, in El-se-car Colliery.

It is not noticed, as far as we have discovered, in M. Adolphe Brongniart's Prodromus; and we almost doubt the propriety of publishing it in this work, because there can be little doubt that it is one of those remains, the identification of which can never lead to any useful result. If, indeed, it were a portion of the stem of a plant, as Mr. Artis supposed, it would have as great a claim to reception among the extinct species of the vegetable kingdom, as any of the others we have published. But if it is, as we hope to show, nothing but the

remains of a mere root, then it will be impossible
to refer it to any class, order, genus, or species,
and, consequently, its recognition will be useless
in the identification of strata ; for it, or what will
not be distinguishable from it, may be expected
in any geological formation of whatever age.

We have, however, thought it as well to admit
a figure of the impression, firstly, for the sake of
explaining what we conceive to be its real nature;
and, secondly, because it seems to throw some
light upon the circumstances under which the coal
measures were formed.

If this fossil were the impression of the stem
and leaves of any plant, there are two points of
structure which would certainly be discoverable
in a perfect specimen. In the first place, the
leaves would be of nearly one size and figure
throughout the branch; and, secondly, they would
be inserted upon the stem with great symmetry
and regularity. As no instance of any departure
from this rule can be adduced among recent plants,
to whatever part of the vegetable kingdom they
may belong, we are justified in considering it,
also, absolute in what regards extinct races ; and,
for physiological reasons, which all botanists understand, the same law is of necessity true of branches;
they also ramify upon a uniform symmetrical plan
from which there can be no real departure. The
subdivisions of this fossil are, on the contrary,
irregular in the highest degree ; no two can be

found precisely alike ; they are of many different sizes ; and they spring from the surface of the central part in a most confused and crowded manner ; nothing even approaching to symmetry, either of form or subdivision, can be detected among them. The fossil, therefore, consists neither of branches nor leaves.

It is among roots, and especially those of water plants that its analogue is to be sought. Irregularity and want of symmetry are the constant characteristics of roots ; and that not only when they have to insinuate themselves among earth, but, also, when they develope in water, or the still more unresisting medium of air. Let, for example, the roots of a melon, growing in water, or of any tree or herb, whose roots have accidentally found their way into a tank, or wet ditch, be compared with this, and their identity will be too striking to be overlooked even by the most careless observer. We, therefore, give the fossil no name ; but merely leave its representation as an explanation of its real nature, for the information of those who had not previously considered the matter.

If, however, its name must be erased from the species of the Fossil Flora, it is not the less interesting in another point of view. Its presence may be considered one of the strong arguments derived from the consideration of organic remains, in favour of the theory that the plants which formed coal were either deposited where they grew, or at

least were not floated from any considerable distance. It is well known that however capable
the stems of plants may be of resisting the action of
water, young roots, and especially those of aquatic plants, are so brittle, that but little violence is
required to break them in pieces ; and if they are
exposed for any considerable time to the action of
a body of agitated water, they would be totally
destroyed. This, on the contrary, is so nearly perfect, that we may reasonably conclude that it had
suffered but little disturbance before it was imbedded in the shale in which its remains have now,
after so many thousand ages, been discovered.

PINNULARIA CAPILLACEA.

From the Leebotwood coal pit, whence specimens
have been communicated by Professor Buckland.

It occurs in small fragments consisting of a
linear central part or axis, from which at regular
distances, on opposite sides, spring capillary appendages divided in a pinnated manner. The
segments of these appendages, exhibit no trace
whatever of leaves, nor in fact any appearance
except that of very narrow dark lines, placed
either in opposition or alternately. At the base
of each opposite pair of appendages the central
part is slightly tumid.

The kind of considerations that lead us to reject
the last subject from the list of fossil species, in-
duces us to add this to the number already described, for it will be found to possess all the characters which we have shewn to indicate stems and
leaves. What we have called the central part we
consider the stem, and the appendages leaves;

leaves, however, which it may be supposed were submersed, if their thinness and want of apparent veins are taken into account.

Had this, instead of the last, been called *Myriophyllites*, nothing could have been objected to the name; for it is so like the submersed part of *Myriophyllum spicatum*, or rather of some of the Indian and South American species of the genus, even to the slight swelling of the stem at the insertion of the leaves, that we do not see how any botanist could prove them to be even different. Nevertheless, as we are quite sensible of the danger of speaking with confidence as to the certainty of such identifications, founded merely upon similarity in external appearance, and especially as the name Myriophyllites has already been applied to a totally different fossil, we prefer coining a new and unexceptionable generic title, which may include any similar remains that shall hereafter be discovered.

From an observation of Count Sternberg in figuring the aquatic leaves of *Myriophyllum*, it appears as if he expected that the fossil genus *Sphenophyllum* might produce such; it is more probable that *Annularia* and *Asterophyllites* consist of the aërial portions of plants whose submersed parts are referable to *Pinnularia*; but this is, in the present state of our knowledge, mere conjecture,

LEPIDODENDRON STERNBERGII.

Lepidodendron Sternbergii. *Supra, vol.* 1. *t.* 4.

The difficulty of determining the species of
Lepidodendron, with anything like accuracy, seems
wholly insurmountable, until we shall have more
positive evidence as to the manner in which the
scars of the leaves were changed in appearance by
the age of a specimen. For this reason we shall
figure whatever illustrative cases we may meet
with, whether they belong to species already described in this work, or not.

Among the plates of Count Sternberg, is one
that represents four states, of what he calls *Lepidodendron dichotomum*, of which one appears to M.
Adolphe Brongniart, altogether different from the
other three. The single figure is supposed to
represent a species already published at tab. 4, of
this work, under the name of L. *Sternbergii ;* and

84

the other three are referred to a doubtful species, thought to be even a distinct genus, called, L. *laricinum ;* to this our L. *dilatatum,* tab. 7, fig. 2, approaches very nearly.

The plant now published, is, we presume, the L. *laricinum.* It differs from L. *Sternbergii,* only in the more truly rhomboidal figure of the scars of the young specimens ; and, perhaps, in the greater size of the leaves. It shows the different states in which portions of the same species may be expected to occur; and, together with an interesting series of specimens which has been put into our hands, by Mr. Prestwich, leads to the opinion that L. *Sternbergii,* and L. *laricinum,* are identical, as Count Sternberg considered them. Fig. *A.* and *C.* are from Hebburn Colliery, and are preserved in the Museum of Sir John Trevelyan, Bart., of Wallington ; at *A,* the leaves are still adhering to the stem; in *C,* they have all fallen away, the scars are altered in appearance, and the dimensions are much augmented. Fig. *B,* is from Colebrook Dale, where it was collected by Mr. Prestwich ; it shows, in a most satisfactory manner, the origin, size, and form of the leaves, which are, it can no longer be doubted, what we call *Lepidophylla.*

LEPIDODENDRON SELAGINOIDES.

Lepidodendron selaginoides. *Supra vol.* 1. *t.* 12.

From the roof of the low main coal seam, Felling Colliery.

This represents L. *selaginoides* in a more characteristic state than the figure before published, in vol. 1. t. 12., and agrees much better with Count Sternberg's plate. It would seem to have been a much branched species, with acute short leaves, closely pressed to the stem; in which circumstance, and its much smaller size, it differs principally from L. *Sternbergii.*

In the specimen before us, the extremities of the branches have all had their bark and leaves stripped off by violence; and from the appearance of the remains of the stripped branches, it seems quite clear that Lepidodendron had a bark, which separated very freely from the woody centre of the stem, just as a modern Silver fir might be deprived of its bark; and hence that, as we have already demonstrated, at tab. 98 and 99, the genus was more nearly related to Coniferæ, than to Lycopodiaceæ; in the latter of which it would be impracticable to separate the bark from the woody axis, without much tearing, or even without destroying the branch itself.

HIPPURITES GIGANTEA.

From the Jarrow Colliery.

The only specimen we have seen of this remark-able plant is that from which our figure was taken. It consists of some fragments of a stem, the joints of which were three or more inches wide, and very nearly three inches long. At the articulations appear the remains of a sheath, divided into a very great number of tapering teeth, which are appa-rently three-quarters of an inch long, and about a line and a half asunder, and present traces of a central rib. The surface of the stem is, in some places, perfectly smooth, without the slightest trace of furrows, or scars; but in other places it presents the appearance of transverse wrinkles.

The stem is pressed quite flat, and evidently was

capable of falling in pieces at the articulations; fragments of several joints being crushed together, and lying one over the other in different directions. Beyond these slight and superficial characters the specimen conveys no information.

Among recent plants we know of nothing to which it can be approximated, except the genera *Equisetum* and *Hippuris*. With the former it agrees in the presence of whorls of tapering leaves, arising from the articulations of a very compressible disarticulating stem; but, on the other hand, all the *Equiseta* have a stem ploughed with deep furrows, and their leaves combined into a sheath much longer than themselves; characters of which no trace can be discovered here.

Hippuris consists of soft-stemmed marsh plants, with narrow verticillate leaves, and the surface of their stem is smooth; but the stem does not readily disarticulate, and is always exceedingly small when compared with such remains as that before us. We find the average of our specimens of *Hippuris vulgaris* to be fourteen inches high, or about seventy times higher than the diameter of their stem; if this fossil were allied to *Hippuris*, and grew in the same proportions, it must have been *nearly eighteen feet in height*.

If verbal distinctions were alone consulted, this plant might be referred to the fossil genus *Asterophyllites*, an heterogenous assemblage of all plants with narrow whorled leaves, seated on a slender stem; but it is incredible that it can have been really allied to such species whatever they were. The whole aspect of the specimens, the different direction of the leaves, and the size of the stem, in the subject of these observations, forbid our referring it to that genus.

A little known plant, called *Phyllotheca Australis*, found in the coal of New South Wales, is described by M. Adolphe Brongniart, as consisting of simple, straight, articulated stems, surrounded at intervals with sheaths pressed close to the stem, as in *Equisetum*, but terminated by long linear leaves, which stand in the place of the short teeth of the sheath of *Equisetum*. We have ascertained, from the examination of specimens, communicated by Professor Buckland, that in some respects M. Brongniart's description of *Phyllotheca* is inaccurate, and that the leaves, instead of springing from the edge of a sheath, arise immediately from the stem, as in the fossil under consideration; so that the two would appear to be nearly allied. But in addition to the whorl of distinct leaves in *Phyllotheca* there is a sheath originating within them, and closely embracing the stem, to which it gives the appearance of the barren shoot of an *Equisetum,* with its whorls of slender branches on the outside of a toothed sheath. Nothing like this remarkable structure occurs in the plant before us.

Upon the whole, we think it indispensible that it should be considered the type of an entirely distinct genus of fossil plants; and as it resembles *Hippuris*, as much as it can be said to resemble anything now living, the name *Hippurites* will, perhaps, be considered not inapplicable.

SPHENOPTERIS ADIANTOIDES.

———

From Jarrow Colliery.

This fine species appears to be undescribed. It approaches to the *Sph. obtusiloba* and *trifoliolata*, in some respects, but it is twice their size, and different in the form of the leaflets.

It was a species with a flexuose, furrowed, slender stalk, whence, at intervals of about three inches, diverged branches, of which the lower were from five to six inches long, and those near the upper end about two inches long. Each of the lower branches was subdivided into branchlets, arising regularly, in a pinnated manner, at intervals of about an inch. The branchlets themselves were pinnated, and bore from three to seven leaflets of a rounded wedge-shaped figure, rather dilated at the upper end, and tapering gradually into a very short slender stalk. Towards the upper end of the leaf, the leaflets, instead of being distinct and forming trifoliate or pinnated branchlets, run together, and become three or five lobed ; and this happens not only near the extremity of the leaf, but also towards the middle and base, giving an irregular and unsymmetrical air to the whole; the circumstance does not occur, we believe, in recent ferns, but we have noticed indications of it in other specimens of fossil ones.

In a specimen of another species now before us, there are two branches that set off from nearly opposite sides of the stalk, a few inches below the point of the leaf ; of these branches, that on the right hand has all its divisions three-lobed, while the divisions of the left hand are pinnated, with from five to seven leaflets.

That this plant was very nearly allied to some of the *Adiantums*, resembling our native *A. Capillus Veneris*, can hardly be doubted ; but, as usual, all attempts at identification have been unsuccessful. The nearest approach to it with which we are acquainted, is in the common *Adiantum* of Chile, which is probably the *A. concinnum* of Humboldt and Bonpland; but that species differs in having longer and slenderer stalks to the leaflets, which are also lobed and crenated.

MEGAPHYTON APPROXIMATUM.

From the roof of the high main coal at Jarrow.
Among the many singular characters that seem
peculiar to the Coal Flora, is that of producing
trees, the branches of which do not grow all round
the stem, as in most modern species, but spring
up in parallel lines, so that the scar of one leaf
is exactly over that which preceded it, and below
that which succeeded. This regular superposition
of leaves, which is known in only a few succulent
plants of the present day, must have been, in the
ages when coal plants flourished, a very common
occurrence; we find it in *Bothrodendron*, in *Ulodendron*, in this genus, and in all the species of *Sigillaria;* a proportion that is remarkably large as compared with the whole vegetation of the same period.
If we exclude ferns, we shall find that about eighty
species of Arborescent Dicotyledonous Coal plants
have been met with, of which nearly half are *Lepi-*

dodendra, or extinct *Coniferæ*, and the remaining half consists entirely of species having the character of their leaves growing in parallel series.

The species now represented is an additional instance of the same kind of structure. Its remains consist of broken stems, which had a dotted, roughish bark, under which appears a surface, ploughed with irregular twisted furrows, which intercept each other without order. On one side of the stem grew leaves, that must have been of very considerable size, if we are to judge by the breadth of the scars they have left behind them. In the middle of the scars are deep discoloured impressions, resembling two parallel horse shoes, *(a, a, a,)* which it may be presumed indicate the figure of the woody system of the leaf stalk. Beyond this nothing can be learned. From such materials, it would be useless to build any theory of the original nature of the plant, especially as we have no recent species with which to compare it. The large size of the impressions, which are thought to indicate the woody system of the leaf-stalks, recals tree ferns to the mind, but neither the arrangement of the leaves, nor the surface of the stem, appears to favour the idea that this can have been even related to the Fern Tribe.

The whole stem of this plant was extracted from the shale, and showed that there were only two rows of scars running up opposite sides of the stem.

MEGAPHYTON DISTANS.

Megaphyton frondosum. *Artis Antidiluv. phytol. t.* 20.

From the shale above the low main coal seam at Felling Colliery.

It was upon such remains as this that Mr. Artis formed the genus Megaphyton, describing it as having an arborescent, simple stem, furrowed longitudinally, with a coarsely fibrous surface. His specimen was larger, and, in some respects, more perfect than this, but the form of the scars of the leaves was less distinctly defined. It is also certain, that the stem is not furrowed, but, like the last, has simply two rows of scars on opposite sides of the stem.

Of the near relation of this species to the last,

whatever the nature of the last may have been, admits of no doubt. It differs, however, specifically in the form of the scars, which do not present the figure of a double horseshoe in the middle, but has only one simple curve (*a, a, a,*) which reaches from one side to the other of the scar.

For what reason Mr. Artis called this *frondosum* he does not state, but as the leaves are unknown, and as they would probably, if discovered, be found to be of a similar nature in both species, we trust we shall be pardoned for altering the specific name.

M. Adolphe Brongniart does not notice the genus *Megaphyton;* we are, therefore, ignorant of his ideas as to its analogy. Until something more shall be discovered concerning it, the character by which it will be known must be the horseshoe figure of the scars, arranged in parallel rows. In a classification of that part of the Coal Flora which contains such things, the genera will be the following;—

* *Leaves or branches, placed one above the other, in parallel rows.*

1. SIGILLARIA. Stem furrowed. Scars of leaves small, round, much narrower than the ridges of the stem.

2. FAVULARIA. Stem furrowed. Scars of leaves small, square, as broad as the ridges of the stem.

3. MEGAPHYTON. Stem not furrowed, dotted.

Scars of leaves very large, of a horse shoe figure, much narrower than the ridges.

4. BOTHRODENDRON. Stem not furrowed, covered with dots. Scars of cones, obliquely oval.

5. ULODENDRON. Stem not furrowed covered with rhomboidal marks. Scars of cones circular.

LEPIDODENDRON ELEGANS.

___

Lepidodendron lycopodioides. *Sternb. vers. fasc.* 2, *p.* 31, *t.* 16, *f.* 1, 2, 4.
Lycopodiolithes elegans. *Ib. Tent. Fl. primord. viii.*
Lepidodendron elegans. *Ad. Brong. Prodr. p.* 85.

___

From Felling Colliery.

Our beautiful specimens of this species consist of remarkably well preserved casts of a large stem and several branches still attached to it. The scars had the acute and regular rhomboidal form of those of *L. Sternbergii,* to which this seems to be nearly allied. It differs in its leaves being much smaller and more delicate, and in the plant having had more slender and graceful shoots. In both species the leaves curve away from the stem, by which circumstance they are essentially distinguished from L. *selaginoides,* whose leaves are closely pressed to the stem.

We are unable to point out any satisfactory marks by which the old stems of *L. Sternbergii* and *elegans* can be distinguished, unless it be the greater breadth of the scars of the former species; a character which we fear will be found too indefinite to be applied with much certainty.

So much has now been said of the genus Lepidodendron in this work, and so very imperfect an idea is, we suspect, entertained of the appearance of those recent coniferous plants to which it is compared, that we shall endeavour to complete the illustration of the genus, as far as it is in our power, by devoting our next plate to the representation of some of those existing species which have the greatest apparent relation to it, and which are unknown in Europe, except in the Herbaria of Botanists. It will be seen how imperfect the ideas of those must be, who have no other notion of coniferous plants than what can be drawn from the pines and firs of European woods and gardens.

PECOPTERIS PROPINQUA.

For the drawing and account of this, of which we have seen no specimen, we are indebted to our indefatigable correspondent, Mr. William Williamson, Jun. He says,

" At first sight, this plant appears to be the same as the *Pecopteris Polypodioides*, figured in a former number, but on closer examination, the outer edges of the segments are found to be undulated ; in the centre of each undulation being placed the sorus, or mass of fructification. From the middle of the segments, veins or nerves strike out, in rather an oblique direction, which are bifurcated ; one point extending to the sorus, and the other in an opposite direction ; both being again bifurcated before they reach the outer margin. Although they vary considerably, I have found this difference in the arrangement of the veins to be a strong distinction between the smooth and undulated edged species ; especially by an examination of the specimens in the choice collection of Dr. Murray, which is always open for the benefit of science. Sometimes one point appears to pass in a single line through the sorus, and the other is twice, or thrice branched, but some part of the nerve *always* extending to the sorus. There is so little of the stem remaining, that I have been unable to discover any peculiar characters; but in the segments, the black carbonaceous matter is well preserved. When a fragment of shale containing one of these plants is split, the black substance forming the sori and midribs, adheres to the opposite side to the one bearing the impression, which occasions the white spots. This specimen was found by my father in Gristhorpe Bay."

120

PECOPTERIS UNDANS.

Of this we have seen no specimens. Mr. Williamson, Jun. has communicated the following memorandum with the drawing we now publish.

" This is one of the most curious plants I have seen found in this neighbourhood. The stem runs in a zigzag manner, and has a line down each side like a Neuropteris. The segments are about two-thirds of an inch long, and rather more than one-eighth in breadth, having a strong midrib which disappears at the apex. In endeavouring to trace the veins, I accidentally destroyed a portion of the black carbonaceous matter; which brought a very singular character to light: *a.* represents the plant as it lay in the stone, shewing the *upper* surface which was curiously undulated; when this part was removed, it left traces of the *under* surface upon the matrix, with two rows of minute sori in the hollow of each undulation, running from the midrib to the sinus of

I 2

the segments, as represented at *fig. c.* This will be the more intelligible if you consider *b.* to be an imaginary view of a horizontal section parallel with the midrib, cutting through three of the undulations, and shewing the position of the sori in the hollows."

Not having seen this plant we are ignorant whether its veins follow the lines of sori, or are otherwise arranged; we therefore place the plant in Pecopteris with which it agrees in habit.

It is from the rich bed of Oolitic plants in Gristhorpe Bay.

SOLENITES MURRAYANA.

Flabellaria viminea. *Phillips Geol. Yorks. with a figure.*

We have been favoured by Dr. Murray, with the following note upon this fossil.

" The plant now sent is from the rich deposit of Gristhorpe Bay, near Scarborough, occurring in the shale of the upper sandstone, belonging to the Oolitic formation; and is so slightly mineralized as to retain *flexibility*, and even in a certain degree *combustibility*. The plant appears to me, most analogous to a Fern, and to the genus Isoetes, to which it is allied by its habit, by the closely matted state of the leaves, by the half flattened structure of those leaves, and by the absence of every trace of leaf-sheaths, or fistular and jointed stems which might have referred it to Gramineæ. Still it can hardly be our Isoetes lacustris.

" By the bye, I have detected in several of our fossil Oolitic vegetables as slightly mineralised in that now sent, some of the vegetable principles, carbon, resin, and tannin."

Upon examining the specimens we found them to consist of very narrow linear leaves, apparently arising from a tufted base, and either adhering loosely to their matrix, as represented at *fig. A*, and leaving a faint impression behind when separated, or collected into firm flexible masses, having little or no adhesion to the mud in which they were imbedded. They were opaque, slightly but not very regularly striated, and taper-pointed, as seen at the magnified figure at *B*. Beyond this striated appearance nothing could be observed of their organization to confirm or invalidate Dr. Murray's suspicion that they were related to Isoetes.

Considering, however, their flexible state, it occurred to us, that if it were possible to separate the tissue from the carbonaceous matter, by some powerful solvent, the transparency of the specimens might be restored and some insight obtained into their anatomical structure. Accordingly, upon plunging them into boiling nitric acid, in a few moments a dark crust peeled away in flakes, and presently the centre part became amber coloured and transparent ; when washed and placed beneath a microscope it was found that all the foreign matter, which had rendered the specimen opaque, was separated, and that the parts were become little less conspicuous than in a fresh specimen. The leaves had become inflated with air, collected

into spaces of unequal size, as shewn at the magnified figures C and D; a transverse section
of them formed an oval, acute at both ends, no
traces of streaks were left, and the sides were evidently composed of prismatical cellular tissue, as
shewn at E, to which internally some soft spongy
matter adhered, which was readily removed with
the end of a dissecting knife, or by frequent brush-
ing with a camel's hair pencil. Not the slightest
trace could be found of veins or of markings in
any way analogous to them.

The recent plants with which this could be compared, besides *Isoetes*, are chiefly *Pilularia*, *Grasses*,
Cyperaceæ, and certain *Rushes*.

From the three latter it differs in the absence of
all trace of veins, which, as they constitute the
hardest part of the tissue, might be expected to be
the longest preserved in a fossil state, and the most
capable of resisting the action of nitric acid; certain species, both of *Isolepis* and of *Juncus*, have
indeed the centre of their leaves filled with a spongy
matter, and some of them have the form which
appears to have existed in this fossil; but in all
cases the exterior coat of their leaf consists of hard
cellular tissue, connecting still harder parallel simple veins. Therefore, it is not with them that we
are to look for an analogy.

The leaves of both *Isoetes* and *Pilularia*, are destitute of veins, and their form, as well as the cellular tissue constituting the shell of their fistular

leaves, is something like those of the fossil. But in the first place, they are divided internally into distinct rows of air-cells, *Isoetes* into four, and *Pilularia* into five or six; secondly, those air-cells are cut off from each other by transverse partitions, which give the leaves, when viewed by transmitted light, their well known barred appearance; nothing of this sort can be found in the fossil, unless the striæ seen on it before exposure to nitric acid, which agree very well with what one finds in *Pilularia*, should be considered as traces of the edges of the rows of air-cells, and the manner in which air collects in the fossil after having been acted upon by the acid, be thought to indicate the existence of transverse partitions. As the partitions in the inside of the leaves of *Isoetes* and *Pilularia*, both those which are parallel with the leaf, and those which are transverse to it, are naturally of a soft spongy nature, they may certainly have been decayed before the plant was finally imbedded, and in that case would be undiscernible now. Of this, however, we must observe there is no evidence.

But supposing that this fossil is admitted as more nearly allied to *Isoetes* and *Pilularia*, than to any thing else now known, which we confess appears to be the fact; it must nevertheless be remarked, that it was distinct as to species at least; for in *Isoetes* the leaves are channelled, or concave and convex, with a sharp keel, and in *Pilularia* they are almost cylindrical, with the upper side deeply grooved, and a thickish edge on each side of the groove, while in the fossil they seem to be what is called ancipital, that is to say, doubly convex with two sharp edges.

We therefore distinguish it as a peculiar genus, for which the name *Solenites* has been suggested, by its fistular structure. Dr. Murray is fully entitled to have it bear his name in addition, in commemoration of his having been both the discoverer of the fossil, and the determiner of its affinity.

A. represents Solenites Murrayana as attached to a mass of mud; *B.* is one of the leaves broken off near the point, and magnified; *C.* a portion of the same, as inflated after having been steeped in boiling nitric acid; *D.* the same, viewed from the edge; and *E.* a highly magnified view of the tissue.

———

Since the above was written, we have received a communication upon the same subject from Mr. Williamson, Jun., who informs us that the plant is common at Gristhorpe, covering the surface of the seams of shale, in every direction. A drawing, which this gentleman has sent us, represents a sort of knob from which the leaves originate. This, so far as it shews any thing, is conformable to the structure of *Isoetes.*

PECOPTERIS LACINIATA.

From the coal mine at Jarrow, where it has only occurred in fragments such as are here represented.

They retain no trace of veins, nor of any other structure that can lead to a comparison of them with other ferns, except their outline; from which we conjecture that the species is closely allied to Mr. Brongniart's *Pecopteris muricata*, from which in fact it appears to differ principally in having the segments of the pinnules usually cut into from 3 to 5 lobes, instead of being entire. The letter-press of *P. muricata* not having yet reached this country, we are not acquainted with the degree of variation to which that species is subject; it may almost be doubted whether this ought to be considered any thing more than a strongly-marked variety of it.

123

SPHENOPTERIS MULTIFIDA.

———

Communicated from the coal measures near
Oldham, by Mr. Francis Looney of Manchester.

It appears to have been a remarkably delicate
Fern, very much like some of the tropical species of
Hymenophyllum or *Trichomanes*; but whether this
is a portion only of a broad leaf, or the principal
part of a small one, it is difficult to judge. From
the slender character of the rachis we should be
disposed to imagine that it was of the latter nature.

The rachis was extremely narrow and slender,
slightly wavy, and triply pinnated, the divisions of
it becoming more and more delicate, till the last
are almost capillary. Each of the first lateral
divisions of the leaf has a broadly ovate tapering
outline, and at its lower part extends beyond those
which are next it; their sub-divisions have the
same outline, and are in like manner so close
together, as to overlie each other near their base ;

the pinnules are deeply pinnatifid, have an ovate figure, and their lobes are cut near their base into five, but near their point into three linear, oblong, acute segments, which are sometimes two-lobed. No trace of veins is left upon any part of the specimen.

The species to which this approaches most nearly are *Sp. elegans*, *gracilis*, and *tenella*. Of these the latter, according to Mrs. Taylor's figure in Brongniart's *Histoire des Végétaux Fossiles*, is only twice pinnated, and the segments of its pinnules are represented as all entire ; we must, however, remark that a comparison of specimens of the two species appears absolutely necessary in order to establish any certain distinction between them. *Sp. elegans* is twice or thrice as large a plant, with obtuse lobes to its pinnules ; and *Sp. gracilis* (from which we cannot distinguish *Sp. Dubuissonis*), has the lobes of the pinnules, both shorter and broader, and only slightly three-toothed.

ASTEROPHYLLITES EQUISETIFORMIS.

Casuarinites equisetiformis, *Schloth. Flora der Vorwelt, t.* 1. *f.* 1. *t.* 2. *f.* 3.

Bornia equisetiformis, *Sternb. Tent. Fl. Prim. p.* 28.

Asterophyllites equisetiformis, *Ad. Brongn. prodr. p.* 159.

First described by Von Schlotheim from the coal measures of Manebach and Mandfleck; recently communicated to us by Mr. Conway from the mines of Blackwood in Monmouthshire.

Its stem appears, from the account of Von Schlotheim to vary in thickness from a line and half to half an inch, according to its age. It must, therefore, have been a plant of considerable size, of which the portions now figured are mere fragments.

We have seen no specimen; but it appears to be of the same nature as *A. longifolia* and *A. galioides* already figured in this work.

Like those species it has been considered analogous to Hippuris, or plants of that nature; but we perceive no evidence of this beyond the verticillate leaves, which prove absolutely nothing, except that the plant was of the Dicotyledonous or Exogenous class.

It is very much to be desired that specimens of this should be found in fructification; for until they have been procured it would be useless to speculate upon its modern analogies.

ZAMIA MACROCEPHALA.

———

For our knowledge of these singularly well preserved remains of what appears to have been the
cone of a *Zamia*, we are indebted to Professor
Henslow, who most obligingly furnished us with
the accompanying drawing, and the following
notes upon it.

"This cone was discovered in cleaning out a
pond about four miles from Deal, on the road to
Canterbury. From the general appearance of the
material of which it is composed, I should think it
must have come originally from the green sandstone formation, and have been accidentally trans-
ported to the spot where it was found. The pos-

sessor, the Rev. C. Yate, Fellow of St. John's College, Cambridge, can give me no further account. Upon comparing it with the figure of a cone of Zamia in Richard's "Coniféres and Cycadées," plate 26, it appears to bear a close resemblance to it in structure, excepting that the scales are longer in the fossil, and curve upwards, in the manner represented in the accompanying sketch.

" I suspect that the cavity which exhibits the internal structure, and shews us so well the arrangement of the seeds, must have existed whilst the specimen was still recent, and that it has not been made since it was found. Perhaps it resulted from the attacks of some preadamite woodpecker. The circumstances which strike me most in this structure, are the slender axis (when compared with Richard's fig.), and the inclination of the seeds consequent on the form of the scales. The diagrams are intended to elucidate the position of the scales upon the cone, according to Braun's views.*

* See an explanation of these views in the forthcoming Report of the British Association for the advancement of science for the year 1833.

The scales on the diagram to the left (above) are so numbered as to indicate the spiral line which winds round the axis, and on which the scales are arranged in succession, and as the thirtieth scale comes vertically over the first, after eleven revolutions of the spiral, the divergence is equal to $\frac{11}{29}$ of a circle ; that is to say, the scale No. 2, is $\frac{11}{29}$ of 360°, or somewhat more than 136°, angular distance, from scale No. 1, and so on of the rest, referring all the coils of the spiral to a plane perpendicular to the axis. The figure below to the right indicates the position of the scales on such an hypothesis."

To these excellent remarks we can have nothing to add except by way of illustration.

The specimen is in light yellowish grey sandstone, which takes a ferruginous appearance when moistened ; it is four inches and three-quarters long, and almost two and a quarter in diameter. At its upper end the scales contract in size, become irregular in outline, and finally surround a small irregular hexagon. At its lower end is a shallow hole rather more than a quarter of an inch in diameter, from which the stalk was pulled out ; allowing for the usual quantity of woody matter forming a sheath round the axis of a cone of this sort, and comparing it with the depth from the surface of the cone to that part of the centre which is actually laid bare, it would appear that the central part or axis, from which the scales arise, was little less than half an inch in diameter.

On one side of the specimen near the base is an opening down to the axis, almost two inches long, and an inch and half wide ; by means of which we obtain a distinct view of the internal structure ; it shews us that the scales curved upwards from the axis, thickening gradually, as represented in Professor Henslow's sketch, towards their point, where they are flat and hexagonal, but not by any means peltate. Near the axis, on the left side, are the cavities left by five ovate seeds, each nearly half an inch long, which have been removed ; their pointed ends are next the axis. On the opposite side are four similar cavities, in one of which is some appearance of the fragment of a seed ; below them is a seed in situ, with a small uneven perforation in its side, and lower still is just visible the thin edge of another seed ; so that these seeds would seem to have had an ovate, somewhat compressed figure, and a prominent edge on each side. In the opinion that this hole was made when the cone was fresh we entirely concur ; but whether by a preadamite woodpecker, squirrel, or mouse, is more than we find evidence to demonstrate.

That it belonged to some Zamia seems to be shewn in every point of its structure, and will be the more apparent if the fossil is compared with the following reduced figures of an American Zamia *(fig. A ;* divided vertically *fig. B),* and of one of the species from Southern Africa, lately named *Encephalartos* by Professor Lehmann *(fig. C).*

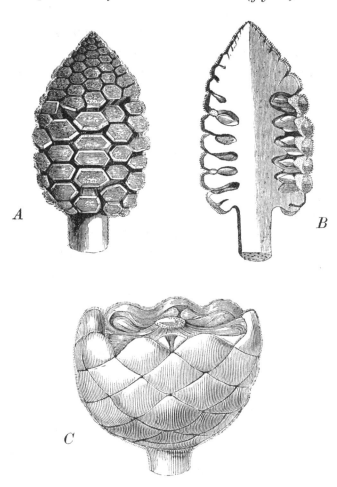

If Professor Lehmann s statement be correct, that
the scales of all the American Zamias have a hexagonal apex, and those of all the African Zamias a
rhomboidal apex, this fossil will then be of the form
now peculiar to the new world ; which is not the
least interesting circumstance connected with it.

To the observation by Professor Henslow, that
the scales are longer, and more curved upwards,
than in the figures given by Richard, we may add
that they also are less distinctly peltate ; but these
circumstances can scarcely be considered to offer
any objection to its being a Zamia, all other points
so nearly coinciding. We possess no materials
whatever for determining how far this may be the
case in some of the many modern species that have
not been figured ; but it can hardly be considered
of more than specific importance.

That this may have belonged to the Greensand
formation is likely enough, considering the great
abundance of the leaves of Cycadeæ in the upper
beds of Oolites, and also that remains of a plant of
similar habit, the Cycadites Nilsoni, has been found
in the lower chalk of Sconen. How different from
its present state must have been that of Europe at
the time when the Greensand was deposited, will
be manifest from the account given by Mr. Ecklon
of the district in which he found the principal part
of the Zamias of South Africa.

They are not met with at Cape Town, where
they would be exposed to the cold winds from the

southern polar regions, but first appear far in the interior of the country, in the land of the Caffers, where the common Cape Flora of Proteas and Heaths is replaced by strikingly different races of plants. They prefer mountainous and wooded, or bushy country, following the ranges of hills, but not straggling into the plains. They are generally met with in rocky places, almost 2000 feet above the level of the sea, higher than the region of Mimosas, and surrounded by bushes of arborescent succulent plants, Rhamneæ, Celastrineæ, and shrubby Leguminous species.

PECOPTERIS WILLIAMSONIS.

Pecopteris Williamsonis. *Ad. Brongn. prodr. p.* 57, *Hist. des Végétaux Fossiles, vol.* 1. *p.* 324, *t.* 110, *f.* 1. 2.

Found not uncommonly in the upper sandstone of the Oolitic formation, near Scarborough.

Mr. Adolphe Brongniart has figured a fine specimen in a barren state ; we are enabled by the kindness of Mr. Dunn of Scarborough to represent it in fructification, a state in which it seems to be not uncommon.

It appears to have been a bipinnated species of moderate size, with a rachis which is often thicker than is usual in most Ferns of the same size. Its pinnæ are narrow, long, and placed on the rachis very obliquely. The pinnules, are oblong, obtuse, curved slightly upward, attached to the petiole by their whole base, and separated from each other by about half their own diameter; in a barren state they have a slender wavy distinct midrib, from which proceed many very oblique veins, which are once or twice dichotomous; in a fertile state, no veins are to be discovered, but the whole of the under surface is covered by a multitude of small projecting circular spots, which it is to be supposed were the sori, or clusters of fructification.

From the complete manner in which the under side of the leaf is covered with fructification, it may be presumed that the elevated circular spots were *thecæ*, and not *indusia* of the nature of those in *Aspidium;* for in recent Ferns it is only the genera with naked thecæ, such as *Acrostichum* in particular, in which the veins and midrib are completely concealed by the fructification; in plants like *Aspidium*, the midrib at least is distinct, however much the veins may be hidden. We therefore conjecture that this *Pecopteris Williamsonis* belonged to the genus *Acrostichum*, to which the disposition of the veins offers no objection.

127

VARIOUS RECENT CONIFERÆ.

We lately promised to give some views of recent Coniferæ, which might serve to illustrate such fossils as Botanists refer to that order, although they have no apparent resemblance to the species with which the European is familiar.

For this purpose we have selected such as are represented in the accompanying plate.

A. Araucaria excelsa, or the Norfolk Island Pine, serves to shew how difficult it is to decide upon the identity of the fossil fragments which we occasionally meet with. The left hand figure is a branch of this plant when it becomes old ; and the right hand figure is a similar branch produced by the plant when it is young ; both taken from the monograph of Mr. Lambert on the genus Pinus. No one could have suspected that such exceedingly different objects as these two could merely be young and old specimens of the same species.

B. Cunninghamia sinensis; illustrates such leaves as *Lepidophyllum;* and may be compared with some of the broad-leaved *Lepidodendra.*

C. Dacrydium cupressinum, a large tree from 50 to 100 feet high, has altogether the appearance of some of the fossils referred to *Lycopodiaceæ.*

D. and *E.* Two undescribed species of *Callitris* from Van Diemen's Land, are not unlike some of the things referred to the genus Fucoides.

OTOPTERIS OBTUSA.

We are indebted to the kindness of Professor Buckland for the drawings from which the accompanying plate has been prepared. The upper fossil is from the Lias at Membury, near Axminster; the lower is from the same formation at Polden Hill, near Bridgewater in Somersetshire. The specimens themselves we have not seen.

It was probably a simply pinnated plant, with a thickish petiole. The leaflets were oblong, obtuse, flat, a little curved forwards into a falcate form, and auricled at their base, on the side nearest the point. They were attached to the petiole by that half of their base, which is not auricled, and were inserted alternately with each other. Midrib they had none; their veins were all of equal size, originating in the base, curving right and left near the sides, running straight in the middle, and forking as much as is necessary to fill the whole leaflet with a dense layer of veins. No structure is visible beyond this.

At first sight it resembles a Fern so closely, that one would scarcely doubt its being one ; but upon a closer examination a circumstance will be detected which will throw some doubt upon the subject. All recent Ferns, with a pinnated structure have, as far as we have observed, either a distinct midrib to eac h leaflet, or, at least, such an arrangement of the veins, as gives the appearance of a midrib ; and we believe it is, in fact, only in *Adiantums* and the *Hymenophyllous* section of recent Ferns, that a midrib is absent, whether the leaf is pinnated or not. But here the arrangement of the veins is such, that not the faintest trace of any thing like a midrib is discernible.

Even in fossil Ferns, or what are so called, it is only in the genus *Odontopteris* that such veins as those of the fossil before us are characteristic ; but in that genus the leaves are bipinnated, and the leaflets grow to the stalk by their whole base, while in this they adhere by only a portion of their base, the anterior half being free and auricled.

Our fossil then is not only doubtful as to its genus, but even as to its affinity, for its veins are not exactly those of Ferns, and its external form is not exactly that of *Odontopteris.*

We find, however, a new red sandstone plant, placed by Adolphe Brongniart in *Neuropteris,* under the name of *N. Dufresnoii,* with which this accords in its veins and mode of division ; but as we cannot consider this species a true *Neuropteris*, for the reasons we have assigned, and as we are now acquainted with at least three distinct plants, which agree in the peculiarities just adverted to, we propose to form them into a new genus, to be called OTOPTERIS, in allusion to the auricle (οὖς) with which the leaflets are always furnished.—*See Tab.* 132.

STROBILITES BUCKLANDII.

From specimens belonging to Miss Bennett, the accompanying drawings were prepared for Dr. Buckland, to whom we are indebted for permission to publish them in this work.

They appear to have been cones, having a slender axis *(a. & b. figs.* 1 *& 2)*, the whole face of which was covered with processes, which at the only remaining surface of the cone have now the appearance of scales. The axis is entirely gone, and the specimens themselves are crushed and broken, as if they had remained in water till they were rotten, and had then been suddenly exposed to some violent action, which broke them in pieces.

On the present surface of the fossil nothing can be traced except the scaly appearance; but it is to be observed, that on both specimens the supposed scales curve *back* from the only end of the cone which is visible ; on which account we conjecture that end to have been the base, for if it had been the apex the scales would rather have converged. At first sight it would seem as if these scales represented the true surface of the cone ; but when we consider the extremely small space which intervenes between the axis (*a.*) and the surface, on the denuded side, and the length of the organs which evidently grow on the opposite side, we find ourselves unable to account for the total disappearance of corresponding organs on the denuded side, except upon the supposition that upon that side the principal part of the cone has been broken away. It would, therefore, appear as if the scales which now remain upon the denuded side, are the bases of bodies, the upper ends of which are left at *c* and *d*.

In the fractured parts, about half way between the axis and the surface of the cone, a number of lozenge-shaped cups (*c. c.*) are visible, with their concavities turned towards the axis ; their margins have a broken appearance, and were apparently continuous with the part which actually grew to the axis. It is to be presumed the cups are the remains of the apex of the cell of a pericarpium.

The parts next the surface of the cone, forming the upper end of the supposed pericarpium, are four-cornered and wedge-shaped, but their points are so buried in the matrix of the fossil that they cannot be made out. At places (*d. No.* 2.) thin plates seem interposed between these wedge-shaped bodies, but we find no evidence to show whether such plates are organic, or mere interpositions of earthy matter.

From the present state of the cones one might imagine that they were originally of an oblong figure; but if our conjecture, that the apparent surface is not the real surface, be well founded, they must have been nearly spherical.

Such is all that we can collect from the remains before us ; scanty as the information is, it seems to shew that the fossil was of a spheroidal figure, and consisted of an axis upon which was planted a number of wedge-shaped, four-cornered, one-celled pericarpia, the upper end of which was solid, and the lower gradually thinned away into a base, which, when the cell was broken off, resembled a scale. Whether real scales were interposed between the pericarpia is uncertain.

It does not appear to us that such information is sufficient to enable a Botanist to determine the affinity of this fossil satisfactorily. That it was not a Fir cone, is rendered probable by the ready separation of the thick four-cornered apex of the pericarpia from the cell, analogous to which we know nothing in *Coniferæ*. For even in *Araucaria*, in which the seed is very large, and terminated by a broad scale, to the base of which it adheres (*See Foss. Fl. t.* 87), there is no such thickening of the upper end as we find in the pericarpium of the fossil; in fact the absence of any distinct trace of a predominance of scales, is not only against its relationship to the Fir Tribe, but also to *Cycadeæ* and *Proteaceæ*.

It is more probable that it was related to some such order as *Pandaneæ* or *Artocarpeæ*. The great objection to the latter is the thickness of the ends of the pericarpia, and the apparent absence of bracteal scales. Such objections do not apply to *Pandaneæ*, the fruit of which is spheroidal, and consists in like manner of pericarpia, often with a thickened wedge-shaped apex, planted upon an axis destitute of bracteal scales, and originally one-celled, although often collected into parcels; and it is to this family of recent plants, that we should be inclined to refer this, if we were obliged to give a positive opinion. But for the present we prefer leaving it in the provisional genus Strobilites, in the hope that the daily multiplying evidence upon this subject will soon enable us to ascertain its nature in a more satisfactory manner.

CYCLOCLADIA MAJOR.

From the roof of the Bensham Coal-seam at Jarrow Colliery.

Like Bothrodendron this plant has branches (?) which readily disarticulated with the stem. All that has been seen of it is in the form of circular depressions about four-tenths of an inch in diameter, arranged in whorls. Its leaves, and the surface of its stem, are quite unknown. What it may have been it would be useless under such circumstances even to conjecture; but as it appears totally distinct as a genus, from all published fossils, we have given it a name by which it may be called. We have another specimen from the coal measures of what seems to be a smaller species *(Cyclocladia minor)*, the diameter of whose scars does not exceed five-twentieths of an inch, but we do not remark any further difference.

SPHENOPTERIS WILLIAMSONIS.

———

Sphenopteris digitata.　*Phillips' Geol. of Yorkshire, p.* 147,
　t. 8, *f.* 6, 7.
Sphenopteris Williamsonis.　*Ad. Brong. Hist. des Veg. Foss.*
　vol. 1, *page* 177, *t.* 49, *fig.* 6, 7, 8.

———

The accompanying plate represents finer specimens of this species, than Mr. Adolphe Brongniart has figured.　The drawings were communicated by our indefatigable correspondent,
Mr. Williamson, Jun., from the Oolitic deposit
at Gristhorpe Bay, near Scarborough, where
the species is rare.

The pinnules are narrowly wedge-shaped,
truncated, often two-lobed, and placed in a
somewhat irregular manner; they often appear
two-parted to their very base, each division
being lobed almost in a fan-shaped manner.

Our upper figure differs a little from the lower in having shorter and more numerously lobed pinnules, which are moreover sometimes confluent; but as they are otherwise extremely similar, are found together, and not unfrequently upon the same stone, we agree with Mr. Williamson, and Adolphe Brongniart, in considering them mere varieties of one species.

Like other *Sphenophylla* this resembles the modern species of Trichomanes, but no one can be named with which it is worth comparing it.

OTOPTERIS ACUMINATA.

From the shale of Gristhorpe, near Scarborough, whence our drawing has been communicated by Mr. Williamson, Jun. The upper and lower figures are from different plants, but appear to represent the same species.

This is so very like *Otopteris obtusa*, figured at plate 128, that it would be superfluous to describe it. In fact, it differs in nothing except its leaflets being much longer, more taper-pointed, and acute, instead of being rounded.

Mr. Williamson has remarked to us that this is in many respects very like *Cyclopteris Beanii* (tab. 44, vol. 1); and upon reconsidering that plant, now that we have become acquainted with this species and *O. obtusa*, we find it necessary to abandon the view we took of the structure of that species, and to consider it a pinnated plant of the same genus with these. It is not improbable that *Otopteris* will have to be reinforced with *Neuropteris Dufresnoii;* but of this we are uncertain, having seen no specimens. In the meanwhile the generic and specific characters of *Otopteris* may be stated thus—

OTOPTERIS.

Leaf pinnated. Leaflets originating obliquely from the side of the leaf-stalk, auricled, attached by about half their base, destitute of all trace of midrib. Veins of equal size, very closely arranged, diverging from their point of origin, and dividing dichotomously at an exceedingly acute angle.

1. *Otopteris obtusa.* Leaflets narrow, oblong, falcate, very obtuse.—From the Lias. Plate cxxviii.

2. *Otopteris acuminata.* Leaflets oblong-lanceolate, acuminate, slightly falcate.—Oolite. Plate cxxxii.

3. *Otopteris Beanii.* Leaflets roundish-oblong, somewhat lozenge-shaped, very unequal sided.—Oolite.

Syn. Cyclopteris Beanii. *Fossil Flora, vol.* 1, *t.* 44.

? 4. *Otopteris Dufresnoii.* Leaflets broadly oblong, obtuse, scarcely falcate, auricled on the lower side.— New Red Sandstone.

Syn. Neuropteris Dufresnoii. *Ad. Brong. Hist. v. f p.* 246, *t.* 74, *f.* 4; *and* 5?

ASTEROPHYLLITES JUBATA.

From the coal measures at Jarrow Colliery.

A thick, blunt, faintly striated, jointed stem, something like that of a Calamite, covered here and there with the remains of a thin carbonaceous layer of what may have been bark, and bearing a multitude of extremely fine thread-like long processes, which it is to be presumed were leaves, are all that we know of this fossil ; which we place in the genus *Asterophyllites*, simply because it accords with the verbal character of that heterogeneous assemblage.

It looks more like a gigantic Equisetum than any thing modern we are acquainted with, but in reality it possesses no character which enables a Botanist to form an opinion about it. All that can be safely said concerning it is that it is a new form in the Flora of the Coal measures.

PECOPTERIS WHITBIENSIS.

P. Whitbiensis. *Ad. Brong. prodr. p. 57. Hist. des Veg. Foss.*
vol. 1. *p.* 321, *t.* 109,*f.* 2, 3, 4.

β. P. Nebbensis, *id. t.* 98. *f.* 3.

" This interesting and beautiful plant was found
in a nodule of argillaceous ironstone, from the
lower shale at Cloughton, near Scarborough.
Like most of our Ferns, the stem, which is the
same thickness in its whole extent, has a depression in its centre, which is also visible on its smaller
branchlets. The leaflets are disposed alternately
in a remarkably regular manner : are of a curved,
falcate form, very acute, and attached by the whole
of the base. Their margins are entire. The midribs are strong ; rising distinctly from the centre
of each pinna, and reaching nearly to the apex of
the leaflets. The veins are forked, springing a
little obliquely from the midrib. The carbonaceous

matter of the stems and branchlets is decomposed, and its situation occupied by the white calcareo-aluminous substance so frequent in the iron nodules. This substance is never found in the shale itself, but invariably in the ironstone, if accompanied by vegetable impressions. I believe it has been described under the head " Scarburgite," and ranked as a mineral. This plant approaches very near to the *Pecopteris insignis* (Fossil Flora, t. 106), and, I think, forms a connecting link between that plant and *P. denticulata* (Neuropteris ligata, *Fossil Flora*, t. 69). It wants the long leaflets of the former, and the dentate ones of the latter, but differs from both in the pinnæ being opposite instead of alternate."

The foregoing extract from a letter sent us with the accompanying drawing, in May last, by Mr. Williamson, jun., contains all that we are able to state concerning the structure of this plant. It is, no doubt, nearly allied to the two species already referred to, but it is essentially distinguished from both by the characters correctly pointed out by Mr. Williamson.

It is more nearly allied to *Pecopteris nebbensis* and *P. Whitbiensis*, especially to the latter. *P. nebbensis*, from the oolitic formation of the island of Bornholm, in the Baltic, as far as can be ascertained from the fragments figured by Brongniart, differs in nothing except its leaflets being rather closer, and obtuse instead of taper-pointed; the veins are represented and described exactly as they are found in this specimen; and it appears to us to be only a slight variety. With regard to *P. Whitbiensis*, figured by Brongniart from the lower oolite of Whitby and Scarborough, the only differences we discover between it and our plant consist in the pinnæ of that species being sometimes alternate, and in the veins of the lower leaflets being twice forked, neither of which was remarked in Mr. Williamson's specimens. To these differences, however, we cannot attach any importance, and we must consider this the same as *P. Whitbiensis*, of which *P. nebbensis* is a variety.

PINUS PRIMÆVA.

——

For the discovery of this we are indebted to
Gilbert Flesher, Esq. of Towcester, who found one
specimen in the stone pits at Burcott Wood, near
that place, and another, which was presented to the
Marquess of Chandos, in Livingstone stone pits.
Dr. Buckland informs us that the formation belongs
to the Inferior Oolite.

This we regard as the nearest approach to the
modern European form of vegetation in the rocks
of such high antiquity as those of the Oolite; for
after a careful examination of it in different directions, we have come to the conclusion that it has
no characters to distinguish it from a modern
Pinus.

It is a cone, which at the time of its deposit had
lost its seeds, and had its scales wide apart, like
those of a Scotch Fir cone, which has been lying

about for some months exposed to weather. Wet earthy matter insinuated itself beneath the scales, filled up all the cavity beneath them, and at the same time, by moistening them, relaxed their tissue, and closed them back again, so as to restore the cone to its original shape. The earthy matter thus formed plates interposed between the scales, and when the latter, which we must suppose were originally decayed at their points, were broken away by the separation of the cone from its bed, projected beyond the scales in the form of a hard earthy border to each scale (*fig. A. a a*).

The specimen we are describing is nine-tenths of an inch long, of an oblong regular figure. It is composed of scales six deep, and six round, the ends of which are rounded, and have a transverse lozenge form ; their surface is finely punctured in consequence of the cellular substance being laid bare by the rotting away of the cuticle and extreme parts. Each scale is dilated at its extremity, and gradually thins away to the lengthened axis (*fig. B.*) of which no trace remains.

The only points in this description at variance with the structure of a recent Pine cone, are firstly, the small size of the fossil : this is botanically of only specific importance ; and secondly, the rounded ends of the scales. In most modern Pines the end of the scales is distinctly and sharply angular ; but *Pinus Strobus* has no angles at the extremity of its scales, and from the worn state of those of the fossil it is most likely that the angles would have crumbled away had there ever been any.

We therefore consider it a true Pinus. That it cannot be referred to any other genus of Coniferæ, to which it bears external resemblance, is easily shewn. *Abies*, which, in the form of the *Larch*, agrees with this in the size of its cones, has scales without thickened extremities. *Taxodium*, the points of whose scales are lozenge-shaped, and which agrees with it in the size of its cones, has no perceptible axis to its fruit, but all its scales spring from a central point. *Voltzia*, which, from its station in the New Red Sandstone, one would naturally compare with it, has all its scales distinctly 3-lobed; and we may add, that this latter circumstance also distinguishes it at once from *Alnus*, whose woody cones, when full ripe, are as large as that of the fossil.

ZAMIA CRASSA.

———

Communicated by Dr. Buckland from the Wealden formation at Yarenland, in the Isle of Wight, where it was found by Mr. John Smith, by whom it was presented to the Oxford Museum, along with a great number of very large bones of *Iguanodon* from the same locality.

The cones appear to have been something more than two inches long, but as their base is lost we cannot be certain of the precise dimensions; now that they are pressed nearly flat they are an inch and half across; they are regularly oblong, and rounded at the extremity. Their surface is covered with deep black, rather irregular, transversely lozenge-shaped scales, which are changed to a brittle carbonaceous matter. Upon cutting through one of these cones, the internal structure, although slightly, is still sufficiently retained to shew that there were numerous seeds lying below the thickened scales at a considerable distance from a thick axis. These are shewn at *a, a, a,* in the lower figure. Nothing can be made of their relation to the scales, except that they are placed immediately below the thickened ends of the latter.

This circumstance disposes of the affinity of the plant which bore these cones to Coniferæ, for in all genera of that order the seeds are next the axis of the cone. And the same point seems to establish their relation to Zamia, to which genus we see no reason why they should not be positively referred : especially considering the existence of other remains of such plants in rocks of a similar age to that of the Wealden clay.

ABIES OBLONGA.

———

Communicated by Dr. Buckland, who believes it to be from the Greensand, near Lyme Regis. It had been washed out of the cliff and rolled to a pebble by the waves on the Dresent shore.

The cone is rather more than two inches and a half long, but was probably longer, for it has been so worn down by constant friction, that its very axis is cut into, and the seeds of the lower part of the cone are laid bare in consequence of the scales that protected them being ground away. Under these circumstances it must not be expected that the external appearance of the fossil is much like what it was when fresh.

Its scales are very broad, rounded, and quite thin at the points; near the axis they are thicker, and apparently consisted of a woody central plate, deeply covered with a corky tissue, which gave way to the pressure of the seeds, forming niches for their reception.

The seeds are so perfectly shewn in a longitudinal section (fig. 2), that not only is their form ascertained to be oval, and their situation at the base of the scales, but in one instance their very embryo may be perceived lying in the midst of albumen. This has been overlooked by our artist, but is plainly visible near the base of one of the halves into which the cone has been cut.

As the position of the seeds near the base of the scales, in connection with other characters, shews this to be *Coniferous*, and as *Abies* is distinguished from *Pinus* by the thinness of the ends of the scales, we have no hesitation about placing this in the former genus, of which it is the second fossil species that has been discovered. To the other, named *A. laricoides* by Adolphe Brongniart, no locality is assigned.

That such a genus should exist in the Greensand, will be by no means improbable if the beds at Titcschen, at Heidelburg, Quedlinburg, and Blankenburg, containing the leaves of Dicotyledonous trees, are correctly referred to that formation.

SPHENOPTERIS CAUDATA.

––––––

Sphenopteris caudata. *Supra, vol.* 1. *t.* 48.

From the shale of Jarrow Colliery.

We trust to be pardoned for republishing this plant now that we have procured tolerably complete specimens ; that which was represented at Plate 48, of our first volume, having been taken from very imperfect fragments.

The impression before us is about a foot long, and comprehends a considerable portion of the upper part either of an entire leaf, or of one of the lateral divisions of a thrice pinnated leaf of considerable size ; one of the pinnæ only and a few fragments, are shewn in our plate.

The pinnæ were set on their rachis at intervals of about an inch and a half; becoming closer towards

the extremity; a line drawn from point to point of
their pinnules would form an ovate-lanceolate acuminate figure, about four inches long, and one inch
and three-quarters wide in the broadest part.

The pinnules are linear-lanceolate, taper-pointed,
pinnatifid, and sessile, gradually shortening towards
the point of the pinna, till the latter becomes
itself pinnatifid only, and finally only serrated.
From their convexity they must have been of a
thick leathery texture.

The lobes of the pinnules are short, ovate, undi-
vided, and obtuse, with a slight depressed rib in the
middle, which vanishes before it reaches the point,
and a very few almost invisible diverging veins;
the former are convex above, and distinctly concave
beneath, where, however, we do not find the slightest
trace of fructification.

We find no published species to which this has a
sufficiently close relation to be worth comparing
with it.

CALAMITES VERTICILLATUS.

———

Professor Phillips has been so obliging as to communicate this with the following note.

"A new species of Calamites from the upper series of the Yorkshire Coal-field. It was found by my friend, the Rev. W. Richardson of Ferrybridge, in the sandstone rock of Hound Hill, near Pontefract, in 1828, and is still in his possession. When we visited the quarry together, it was interesting to remark, that though in general the Ferns and other delicate plants are rarely found in open-grained gritstones, fronds of Pecopteris, stems of Halonia, fruits reminding us at least of some of the Palmæ, Lepidodendra, Calamites, and other plants, were entombed together in this rock."

It is different from any species that has yet been met with, on account of its distinct whorls of large deep scars, which represent the points of

o 2

attachment of so many branches. This discovery will probably be found to assist us very much in forming an opinion upon the real nature of this singular genus, whenever we shall succeed in finding a clue to the right understanding of what such puzzles as *Calamites, Sigillaria,* and *Stigmaria* really were.

140

CAULOPTERIS PHILLIPSII.

——

For a drawing of this very distinct species of Tree Fern stem, we are indebted to Professor Phillips, who communicated it with the following note.

" This is the plaster cast of a fossil stem from Camerton Colliery in Somersetshire, where the specimen was, I believe, found in the year 1800. It was, I think, in the possession of the late C. J. Harford, Esq. a friend of the late Rev. J. Townsend of Pewsey (author of a well known geological work, embodying many of Mr. W. Smith's early views) and of the late Rev. Benjamin Richardson of Farley, in whose collection this plaster cast was preserved. It was given to me by Mrs. Richardson in 1833. I consider it to be the stem of a Tree Fern, different probably from any yet published. I may remark that I have never seen any fossil stem which appeared to possess the character of a Tree Fern

from any British Coal-field except that of Somersetshire.

" No particular markings are observable in the
cast between the cicatrices, but the intervening
spaces appear nearly smooth. The cicatricial
markings are not all similar, and I find on some
recent Tree Ferns considerable variation in this
respect, arising apparently from the singular rup-
ture of the vessels, &c.

" The cast includes probably the greater part of
the breadth of the plant; it is of an oval figure in
the cross section, in consequence of compression."

It is obviously distinct from *C. primæva*, figured
at tab. 42, and these together with the little *Cau-
lopteris gracilis*, published at tab. 141 of the present number, form the only Tree Fern stems we yet
have met with in the Coal Measures.

CAULOPTERIS GRACILIS.

An extremely rare fossil, belonging to the Ketley, Coal-field. The only specimen we have seen was communicated to us by Mr. Prestwich, Jun., "from the shale of the Pinny Iron-stone measure, at the Hay-pits, Madeley; it was found associated with large quantities of marine shells." It also exists in the collection of Mr. Austin of Madeley.

Our specimen is a hollow cylinder, marked internally with deep and distinct longitudinal fissures, about half an inch long, alternating with each other, and piercing the whole thickness of the cylinder, so that where the latter is broken across it is separated into lobes of unequal width, as is shewn in our figure. Externally the surface is covered irregularly with elevated lines, which appear to be the remains of fibres that were attached firmly to the surface; it is also pierced here and there with fissures which communicate with the inside.

We know of nothing among recent plants to which this can be compared except a slender Fern-stem; with which we are disposed to identify it, notwithstanding the absence of the scars of leaves, and its fibrous surface.

In all Tree Ferns the scars disappear towards the lower part of the stem, where their place is occupied by a layer of entangled fibres; so that this, if a Fern stem, must have been the lower end of one.

The cylinder of which the trunk of a Tree Fern consists, is composed of a number of irregular lobes, which are the bases of the leaves, adhering to each other by their sides; in this specimen the fissures may be considered the lines of contact of such bases. We do not, however, know any recent Fern in which the bases of the leaves adhere to each other so slightly as to leave passages between them; but in *Dicksonia arborea*, the internal furrows are so deep that this nearly happens.

Each base of a Fern leaf, consists of an external coating of a hard texture, and of a softer substance in which a number of sinuous plates are arranged. It often happens that the soft substance shrinks away from the hard outer case, thus leaving a space between the two; precisely the same thing seems to have happened in this fossil (see fig. 2).

Upon the whole we regard it as tolerably certain that this was the base of a slender Fern-stem; and upon this supposition we especially recommend it to the consideration of those who occupy themselves with the study of the economy of recent Fern-trunks. If we are not greatly mistaken, it is calculated to throw no inconsiderable degree of light upon what has hitherto been a very obscure subject.

TRIGONOCARPUM OVATUM.

———

Communicated by Mr. Prestwich, Jun., from the Pinny Iron-stone measure at Ketley : it is now in the collection of Mr. Austin of Madeley.

The existence of Palms at the time of the Coal measures has always been insisted upon as one of the many proofs that the Vegetation of the Coal æra was tropical; but this, like the arguments derived from the supposed existence of Tree Fern stems, has long been exposed to objections which are not easily answered. We have shewn, at table 42 of the first volume of this work, that up to the time when that article made its appearance, there had not been a single genuine Tree Fern stem described from the old Coal of any part of the world ; now, with what are published in our present number, the existence of three English species will have been demonstrated. So with Palms; no one has yet seen Palm-*wood* in the Coal measures, only three kinds of *leaves* have been referred to this class, and of those, one, the *Flabellaria Borassifolia,* is probably not a Palm at all; while the other two, both belonging to the genus *Nöggerathia,* are by no means so clearly proved to be Palms that a question could not be raised about them, especially in the absence of proof of the existence of other species; and finally, doubts have been expressed by Adolphe Brongniart (*Prod. p.* 120), whether the fossil Coal fruits, supposed to belong to Palms, were not in fact something else.

Under these circumstances, we think we shall be rendering good service to Geology if we can succeed in producing tolerably good evidence, in two more cases, of the existence of Palms in this country at the time when the Coal was deposited, and a third which is supported upon testimony which the most scrupulous Botanist cannot gainsay.

The first to which we have to call attention is the subject of this article (t. 142, f. A). This was an ovate fruit, of the exact size shewn in our drawing, originally covered with a thin coat, which now remains in the form of a thin broken carbonaceous crust; below this coat was a thick shell marked with three projecting ribs, and within the shell was a single seed which seems to have stood erect in the cavity; all this is visible in our specimen, in consequence of the shell having been broken through from the apex, so as to lay bare the seed. The latter seems to have been soft at the time when it was converted into ironstone, for there is a distinct trace of a deep depression in two places, just at the point where the shell is fractured. No trace of calyx, or of any other body is discernible externally.

Now all this is exactly what would be seen in many Palms, which have in like manner a three-ribbed fruit containing a single seed within a thick shell, and if their seed were decayed, its sides would give way just as has happened here, in consequence of its being hollow like the Cocoa-nut; such a Palm is the common Chilian *Micrococos*, which is so commonly sold in the market of Valparaiso. Supposing the apex of such a Palm could be laid bare by a fracture of the shell, as has occurred in our fossil, a number of veins would be seen passing downwards from the apex towards the base; traces of such a structure are distinctly visible here, only they are scarcely elevated above the surface of the seed, which may have been caused by the decay of the latter.

No doubt this is nearly allied to *Palmacites dubius* of Sternberg, which Brongniart calls Trigonocarpum dubium, but that species is both rounder and smaller.

POACITES COCOINA.

Obligingly communicated to us from the Lancashire Coal-field, by Dr. Black of Bolton.

The only two species we have seen of this, are the present, and another from Bideford, in Devonshire, among some vegetable fossils, collected by Mr. De la Beche, and in both the two parts of which the species consisted were placed obliquely with respect to each other, as is represented in the drawing ; the one half having convex veins, and therefore shewing the lower surface, while the other half is proved, by its concave veins, to have been the upper surface. It is evident that they were applied to each other face to face, and one would think that their relative position was caused by their having been doubled down upon each other.

From the great breadth of this leaf, and its apparent length, it could scarcely have been any thing except the leaf of some pinnated Palm, whose pinnæ are of considerable width, as in many species of

Cocos; at least we know of no other monocotyledonous leaf with which it can be compared.

Supposing this analogy to be a just one, it is not impossible that the position of the two faces, which seems to be caused by the leaf being doubled up, may be owing to the original structure of the leaf itself. For if it is the remains of a simply pinnated leaf, the under side might belong to one pinna, and the upper to another, pressed against each other in consequence of the leaf being folded up. And this we are the more inclined to suspect may be the case, in consequence of both the specimens we have seen, from distant localities, being in just the same state, a circumstance which would hardly have occurred if the doubling of the leaf were accidental.

This we regard then as a second new instance of the existence of Palms in the Coal measures.

142 C

TRIGONOCARPUM NÖGGERATHI.

———

Trigonocarpum Nöggerathi. *Ad. Brong. Prodr. p.* 137.
Palmacites Nöggerathi. *Sternb. t. 55. f.* 6, 7.

———

Whatever opinion may be held of the relation
of the last two fossils to Palms, there cannot be the
slightest as to this, for which we are also indebted
to Dr. Black.   It occurs in considerable quantity
here and there, imbedded in sandstone, as if it had
originally grown in large clusters : as was in all
probability the case.   We regard this as by far the
most interesting fruit yet met with in the Coal
measures.

It is possibly to this that Adolphe Brongniart
alludes, when speaking of two or three species of
hexagonal fruit, found in the Coal, which he considers cannot be Palm fruits, " because in all the
genera of this family, when the fruit is symmetrical

it consists of three parts and not of six." Upon this we must remark, that although a six-sided figure is not common in Palms, yet it exists in *Diplothemium maritimum;* and that moreover this may be proved to be a Palm upon the clearest evidence.

The principal part of what we have examined consists of specimens of an ash grey colour, almost exactly oval, but more acute at one end than the other, and marked with three acute and three obtuse ribs, of which the latter are but little elevated. Fig. 1, represents a side view of one of them ; 2, the base, and 3, the apex : in this there is nothing that can be called evidence. But upon fracturing a mass of sandstone, in which great numbers of fruits were imbedded, we were so fortunate as to obtain a distinct view of the internal structure, as represented at fig. 4 ; from which it appears that the fossil in its ordinary state, is an interior part divested of a fleshy covering.

It consisted originally of a soft coat (fig. 4, *a.*), and was blunt at the apex, but tapered into a stalk (fig. 4, *e.*) at the base. Within this was another covering (fig. 4. *b.*), which enclosed a single seed. In the specimen the lower end of the seed was depressed as if it had been softened ; in the centre (fig. 4, *c.*) it had a small round depression ; and a number of veins passed downwards from its apex, losing themselves near the middle of the seed.

Now all this is so completely the structure of a

Palm, that there can be no doubt whatever that this fossil was the fruit of a plant of that kind ; indeed the depression in the centre (fig. 4. *c.*), which indicates the seat of the embryo, and the raphe so rich in veins, are to be found combined in no other plants.

In fact, let any one compare it with a Date-fruit, and it will be impossible not to recognise the great similarity in organization.

It is, however, very remarkable in this fossil, that although it has apparently the drupaceous structure of such fruits as the Cocoa-nuts, yet it has no pore provided for the escape of the embryo. It is impossible for so small and weak an organ as the embryo of a Palm to force its way through so hard and thick a covering as a Cocoa-nut shell, and, consequently, nature thins the shell over against the embryo, in order to enable the root of the latter to find its way into the earth ; this contrivance is seen on a Cocoa-nut shell, in the form of the three well known black spots at the end. It is to be expected that some trace of this contrivance would be discernible here; but as that is not the case we must suppose that the second coat of the fruit, which answers to the stone, was in this instance soft enough to render such a provision as an embryo-pore unnecessary. Upon this supposition it will have belonged to a genus essentially distinct from any at present known.

CYCADEOIDEA PYGMÆA.

———

Communicated by Professor Buckland, from the lias at Lyme Regis. The specimen belongs to Miss Philpotts.

At first sight this might be taken for the cone of some tree ; but the irregularity of its figure, and of the arrangement of the scars upon its surface, together with the appearance of a large tubercle on one side, will alone throw doubt upon the correctness of such an opinion ; and this doubt is increased by the absence of all trace of seeds in a polished vertical section. When cut through from the apex to the base, nothing can be seen except the bases of blunt scales, planted perpendicularly upon a thick and solid centre.

In fact, we entertain little doubt that instead of a cone, we are to consider it as the stem of a small species of Zamia, analogous to those productions in the Isle of Portland, the real nature of which Professor Buckland has so satisfactorily elucidated in the Transactions of the Geological Society. Upon this supposition the tubercle near the middle will be a rudimentary branch, and all the irregularity of form and arrangement in the spaces which cover the surface, especially near the base, will be consistent with what we should find in nature.

Our figure is taken from a beautiful drawing by Mr. Sowerby, for which we are indebted to the liberality of Professor Buckland.

144

PHLEBOPTERIS CONTIGUA.

———

This genus is figured in the 83rd plate of Brongniart's History of Fossil Plants, but the letter press has not yet reached us. It appears to be distinctly characterized by the presence, next the midrib, of a row of areolæ, the upper edge of which is either oblique or parallel with the midrib, on which the simple or dichotomous veins are planted almost perpendicularly.

As a species this is obviously distinguished from Brongniart's plant, by its pinnæ being so close together as to touch each other at the edges, and much wider, while their costal areolæ are oblique instead of semi-hexagons.

It was found in Iron nodules in the Oolitic formation of Gristhorpe Bay near Scarborough, and was communicated by our excellent correspondent Mr. Williämson, Jun., with the following note.

" The central stem has tapered very rapidly,
and is rather strongly striated.   The greater part
of it, however (as well as the central nerve of the
leaflets), is decomposed as usual.   The leaflets are
alternate, slightly curved upwards, about one inch
and a half long, terminating in an obtuse apex.
The divisions do not quite descend to the central
stem, but their place is occupied by a remarkable
arrangement of the nerves, which will be better
understood by the magnified drawing than by my
describing it.   The small spaces on each side of
the main nerve are rather irregularly formed, sometimes opposite and in others alternate, but more
frequently the former, so as to shew a string of
curious heart-shaped appearances in the centre of
each leaflet.   The nervures are sometimes divided
near the margin ;   about every second and third.
I cannot discover any traces of the sori Brongniart
mentions : they either do not exist in our specimen,
or are very minute, and on the under side of the
leaf, so as to be invisible.   This is the only specimen I have seen : we have another which differs
from this, in the nerves not dichotomizing at the
margin."

145

PECOPTERIS MANTELLI.

Pecopteris Mantelli. *Ad. Brongn. Hist. des Veg. foss. v.* 1, 278, *t.* 83, *f.* 3, 4.

For this we are indebted to Mr. Conway of the Pontnewydd works, who obligingly communicated an excellent drawing of it with the following note.

After noticing its great resemblance in some respects to *Pecopteris heterophylla, tab.* 38 of this work, this gentleman remarks, "that the difference between the two will be found sufficiently great to form them into distinct species. The pinnæ of this plant are much longer, and neither so much tapered nor so acutely pointed as in *P. heterophylla;* but the most remarkable difference consists in the terminal leaflets, A and B, which give this specimen quite a distinct character, and must have produced a very graceful habit in the living plant. The specimen is from the Coal Mines of the British Iron Company at Abersychan in Monmouthshire, and is the only one I have ever seen".

It does not appear to differ from *P. Mantelli*, of which Adolphe Brongniart has given a figure, from a specimen without the terminal pinnæ, communicated to him by Mr. Mantell, from the Newcastle Coal measures. That learned Geologist compares it with the common *Pecopteris lonchitica*; from which it is obviously to be distinguished by its very narrow and obtuse pinnæ, independently of the long terminal one. Like *Pecopteris heterophylla* it represents an extinct form of *Pteris*, of the nature of *Pteris caudata* and *aquilina*. Adolphe Brongniart regards it as intermediate between *Pteris caudata* and *arachnoidea*, two West Indian species.

146

SPHENOPTERIS CONWAYI.

———

For this also we are indebted to the same intelligent correspondent who furnished us with the last subject.

Mr. Conway observes, that "there appears to have been a very peculiar character belonging to this Fern. It seems to have been coriaceous and very thick, so as to give to the whole plant somewhat of a tuberculated appearance. Each portion of the leaflet appears as if formed of a separate globule, and the globules seem, by compression, to have been squeezed into each other, and thus to form one mass. This may possibly arise from the plants being in fructification. When the impression of the under·side is left in the shale it is in very deep indentations; and, if these indentations are the impressions left by the sori, then they must have been arranged somewhat in the same manner as those on the *Aspidium Filix mas* of the present day. The pinnules are attached to the rachis by the whole of their base, and the veins radiate, as it were, from the base of each apparent tubercle of which the frond is composed, without any division or branching. This I have endeavoured to represent in a magnified portion. The only specimens I have seen are from Risca, in this county, and there, I understand, it is a common fossil."

Not having seen a specimen we are able to add but little to the foregoing remarks.

The fossil obviously belongs to the set of *Sphenopteris*, consisting of *S. Höninghausi, rigida, trifoliolata*, and *obtusiloba*, which Adolphe Brongniart justly compares with the larger species of the modern genus *Cheilanthes*. They all are Coal measure plants, having that character of convexity in the lobes of the pinnules, which Mr. Conway justly describes as giving the plant a tubercular appearance. If, however, they were really related to *Cheilanthes*, it is to be remarked that this convexity was not owing to the pressure of a large central sorus beneath each lobe, but to the curving backwards of the edges of the lobes so as to cover the narrow marginal sori.

From the species described by Adolphe Brongniart, this differs essentially in the pinnules being seated close upon the rachis, and touching each other, so that to the naked eye the pinnæ look as if they were regularly pinnatifid with very short acute lobes. Each of these supposed lobes is in reality a pinnule, consisting of three or five lobes, of which the lowest are much the largest, and the terminal one rather narrower and longer than the intermediate ones, if there is any of the latter present.

We have named it in compliment to the gentleman who has so obligingly communicated it to us ; as a slight acknowledgment of the value we attach to his investigations of the highly interesting Coal flora of South Wales.

From the appearance of this specimen it may be conjectured that it was a Tree Fern ; for although there may be some doubt whether the lateral ramifications are in all cases actually attached to the central rachis, yet the general relation borne to each other by the parts as they lie imbedded in the shale, is such as to render it highly probable that they all once belonged to each other. In this case the species would not be very widely different from the *Cheilanthes arborescens,* a Tree Fern which now inhabits the New Hebrides.

147

SPHENOPTERIS POLYPHYLLA.

———

Communicated by Mr. Murchison,* from the coal of the Titterstone Clee, in Shropshire, where it was found by Mr. Lewis.

* This, with many others, some of which form a part of the present Number, was collected by Mr. Murchison, during his recent geological surveys of Salop, Hereford, and the adjoining counties. These plants are from the Knowlsbury coal-field, a small elliptical basin, situated at the south-western termination of the carboniferous tract of the Titterstone Clee Hills. They occur chiefly in the roof of the great coal, and gutter coal, and also in the concretions of iron-stone. It is important to remark, that in a Memoir lately read before the Geological Society, Mr. Murchison has shown, that the Clee Hill coal-measures, as well as those of Coal-brook Dale, of the Wyne Forest, and of Oswestry, are all of *older date* than those of the Shrewsbury field. The latter containing a fresh-water lime-stone, and passing upwards into the base of the newer red sandstone, is proved to be the youngest of these carboniferous zones. It is from a portion of this Shrewsbury coal-field (Le Botwood), that we formerly published the specimens of Neuropteris cordata, Odontopteris obtusa, and Cyperites bicarinata, figured in our first volume, which plants Mr. Murchison has discovered in various parts of the same field, associated with Pecopteris lonchitica.

It is a very distinct species, allied to *Sph. obtu-siloba*, but decidedly different in the lengthened form of the central piece in all the three-lobed segments of its leaves.

The leaves were bipinnated at least, and possibly more frequently divided; both the principal and secondary pinnæ were so closely placed that the lobes over-lapped each other. The segments of the pinnæ had an ovate, or somewhat heart-shaped figure, and were divided into from three to five lobes. When the lobes were five, the terminal one was not much longer than the others, but they all had a rounded termination, and the lateral were sometimes split; when the lobes were only three, the terminal one was always much longer than the two lateral ones, which near the point of the pinnæ became mere auricles and finally disappeared. The veins were wide apart and almost always forked.

148

SPHENOPTERIS SERRATA.

Discovered in the lower sandstone and shale of the Oolitic series at Cloughton Wyke, near Scarborough; for the drawing we are obliged to Mr. Williamson, jun.

The leaves were bipinnated, with all their divisions regularly alternate. The pinnæ were five or six inches long, and consisted of about twenty-four pairs of narrow, very regularly, and deeply serrated lobes, which gradually tapered to a narrow but not acuminated point. Each division of its lobes is represented by Mr. Williamson as having a set of very delicate veins passing towards the point, and sending off simple veinlets obliquely and laterally.

Only one specimen has been met with in a coarse-grained sandstone, much impregnated with iron.

From nearly the whole of the lobes of this plant having had their ends abruptly broken off, it is not improbable that it had been lying for some time in troubled water before it was deposited, for we have remarked the same circumstance occur in recent ferns which have accidentally fallen into a large piece of water, provided they have their lobes serrated in the same degree as this species.

Polypodium and *Aspidium* are two modern genera, both of which contain species analogous to this, but we have not succeeded in identifying it with any recent plant.

We have no fossil *Sphenopteris* with which it is at all necessary to compare this.

149

SIGILLARIA MURCHISONI.

From the Knowlsbury Coal-field, whence it was brought by Mr. Murchison, after whom we have named it.

There is no species which has yet been published for which this can be mistaken. It is most like *S. oculata*, but is quite distinct from it and all others on account of the singular markings of its surface.

The only specimen we have seen consists of six broken elevated ribs with concave spaces, about four lines wide, between them. The scars are exactly the form of those in *Sig. oculata*, and have a double or triple point of communication beyond their centre. The surface between the ribs is very sharply and distinctly marked with broken wrinkles which form curves connecting the ribs; these curves are by no means uniform or regular, but are placed at unequal distances and often anastomoze.

150

OTOPTERIS? DUBIA.

The only specimen of this curious plant that has yet been met with is one which was procured by Mr. Murchison, from the sandstone of the Knowlsbury coal-field.

If really an *Otopteris* it will be extremely interesting, as being the first species of its genus that has occurred in the coal formation. Hitherto it has been supposed confined to the Oolitic series, if we except a doubtful plant from the New Red Sandstone. (See page 142 of this volume).

This specimen is so very like *O. obtusa* (tab. 128) in size and general appearance, that although it is essentially distinguished by its leaflets being narrowed to their base, instead of being broad and auricled, we are led to suppose it may belong to the same genus. There would be no doubt indeed of the matter, if the embedded leaflets were all decidedly upon the same plane, for then we should be sure that it was really a pinnated leaf; but the leaflets are so irregularly imbedded in the sandstone, some being visible upon fractures of the surface considerably lower than others, that we cannot avoid entertaining a suspicion that the leaflets, or rather leaves, as in that case they would be, were either whorled or placed all round a slender stem. Should this be so, the plant would then be a new species of either *Sphenophyllum* or *Trizygia*, both of which are genera confined to the Coal-measures; and this is perhaps the more probable supposition.

As far as we can make out, the ends of the leaflets were rounded as we have represented them, but we cannot be sure that the margin has not been broken away.

151

SPHENOPTERIS MACILENTA.

Found in the coal mines at Risca in Monmouth-shire; and communicated to us by Mr. Conway.

The only specimen which has yet occurred, and which is that now represented, is a very perfect impression of a pinnated leaf, the pinnæ of which are deeply pinnatifid at the base, but with con-fluent lobes at the apex. The lobes appear to have been very thin and delicate after the manner of recent Adiantums: the lowermost on each pinna were roundish, contracted at their base into a very short stalk, and pretty regularly three-lobed. As they approach the apex the lobes lose all trace of a stalk, become entire, and at last are confluent into a tapering pinnatifid extremity. The veins are so delicate, or have been so imperfectly pre-served, as to be scarcely visible when they approach the margin of the lobes; nearer the base they are more distinct, and spread regularly from their origin, bifurcating as the lobe dilates.

This is nearly allied to *Sphenopteris adiantoides*, already figured at t. 115 of this work ; but it differs essentially from that species in the tapering form of its pinnæ, and in the division of its lobes. Whether it was pinnated or bipinnated the specimen does not enable us to determine. It is also closely related to *Sp. latifolia*, t. 156, but was a plant of a much larger size in all its parts.

152

LEPIDOPHYLLUM TRINERVE.

Sent from the Coal-measures of Blackwoodia, Monmouthshire, by our obliging correspondent Mr. Conway.

Figs. 1 and 2 represent the fossil as it has occurred ; fig. 3 is a sketch by Mr. Conway, of the manner in which he conjectures the leaves to have been formed. He observes at the base of each leaf a kind of plicature, as if its substance was a little wrinkled, and from the nature of the plaits it would seem that the leaf-stalk had been of a thick leathery texture.

This agrees well enough with the structure of *Araucaria*, and the close contact in which it is obvious from fig. 2, that the leaves must have grown is further confirmatory of the opinion that

Lepidophylla are the leaves of some plants very similar in manner of growth to the South American species of that genus.

The three strong veins in the leaves are characteristic of this species.

153

PECOPTERIS LONCHITICA.

Planta diluviana epiphyllospermos in saxo dimidiato convexo-plano in profunditate ingenti reperta in fodinis ferri prope Newcastle Northumbriæ. *Scheuch. herb. diluv. p. 74. t. 1. f. 4.*

Filicites lonchiticus. *Schloth. Petrefakt, p. 411. Flora der vorwelt, p. 54. t. 11. f. 22.*

Alethopteris lonchitidis et vulgatior. *Sternb. Fl. der vorw. fasc. 4. p. xxi. t. 53. f. 2.*

Pecopteris blechnoides. *Ad. Brong. Prodr. p. 56.*

Pecopteris lonchitica. *Id. Prodr. p. 57. Hist. des Veg. Foss. p. 274. t. 84. f. 1—7. t. 128.*

One of the commonest of the plants of the old coal formation, occurring in great numbers in various mines of France, Bohemia, Silesia, and England. It has lately been met with in great numbers by Mr. De la Beche in coal at Bideford in Devonshire.

The fragments in which it is found being often from different parts of a leaf, are sometimes so different in appearance as to have led to the formation of several spurious species. When the pinnules are decurrent it is the *Alethopteris lonchitidis* of Sternberg, when rounded at the base it is the *Alethopteris vulgatior* of the same author. Adolphe Brongniart originally separated it into *Pecopteris lonchitica* and *blechnoides*, but afterwards combined them; and we do not see on what character his *P. Serlii* from the Bath coal-field is to be distinguished.

From a comparison of various specimens in different states, it is to be gathered that this plant was a bipinnated fern with leaves about the size of those of the common Brake, or something larger. The lobes were long, narrow, and usually decurrent, contracting, however, at the base towards the lower part of the pinnæ. In some cases they were acute, in others acuminated, and occasionally they were rather obtuse. Towards the end of the pinnæ they became confluent, diminished very much in size, and at last ended in a long lobe, resembling those at the base of the pinnæ both in size and form. In all cases they were strongly marked with a midrib, on which were placed almost perpendicularly a number of close fine veins which are usually simple but sometimes forked.

In all the specimens we have examined the lobes of the leaves have uniformly been convex, and sometimes in a remarkable degree; this circumstance, which shews that they were originally of a thick texture, taken together with the general resemblance of this plant to certain species of *Pteris*, especially to *P. aquilina*, and several Indian kinds, has led to the suspicion that it must have been in fact a species of that genus ; and Adolphe Brongniart has stated that it is most nearly allied to *Pt. caudata*, a West Indian plant. It must, however, be observed, that the veins in that species and in all those of the same division of the genus, are much more distant and forked than in this, which, if a Pteris at all, we should consider more nearly allied to some of the simply pinnated *Pterides*, notwithstanding its greater degree of division. It is, however, entirely different from all the recent species of which we have any know-ledge, and in fact so nearly agrees with several *Blechna* in its veins, especially *B. orientale*, that we are by no means sure that the weight of evidence is not in favour of its being a *Blechnum* rather than a *Pteris*.

154

PECOPTERIS DENTATA.

———

P. dentata, *Ad. Brongn. prodr.* 58. *Histoire des vég. foss.* *vol.* 1. *t.* 124.

———

From a coarse micaceous shale in the Newcastle Coal-field.

The portion here represented was the upper part of a pinna of a tripinnated plant, which must have been of considerable size; judging from a noble specimen figured by Adolphe Brongniart probably arborescent. Each pinnule was on an average an inch and a half long, and the distance between the setting on of the pinnules was about four lines. The lobes were placed close together, and were about two and a half lines long; at the base they were slightly united, at their points they were rather acute, their sides were nearly parallel and crenelled ; they were traversed by a midrib, which reached to their apex, and gave off obliquely a number of distant forked veins.

This differs from *P. pennæformis*, another Coal-measure plant, in almost nothing except the crenelling of the lobes of the leaves, as far as we have any means of judging ; but as the letter-press of Adolphe Brongniart's work, containing the description of this has not yet appeared, we are ignorant of the motives he has had for separating them.

OTOPTERIS CUNEATA.

—

We are acquainted with this remarkable little plant only from the accompanying drawing, and a description with which we have been favoured by Mr. Williamson, jun.

It was discovered at Gristhorpe Bay near Scarborough, and is supposed by Mr. Williamson to have been a fern belonging to the genus *Glossopteris* and having its leaflets both springing from a common point. The veins are described as being twice or thrice forked between the setting on of the leaf, and the margin ; the leaflets, in the only two specimens that have been met with, appear to originate from the apex of a short common stalk, and were roundish-oblong, with something of a wedge-shaped outline.

As the leaflets have no midrib, but are mere membranous expansions, traversed by veins radiating from the base, and branching at regular intervals, so as to fill up the parenchyma, it is not possible to refer this plant to *Glossopteris*, neither is it very easy, in the absence of a greater number of specimens, to know in what other genus to station it. We are not sure whether it really did consist of only one pair of leaflets, and we do not know whether the stalk is all that the plant ever had, or whether it is not a part of something very much branched. It is, however, most probable that it was allied to *Otopteris Beanii*, and the other species of that very distinctly marked group, and it is thither that we think it safest to refer it.

156

SPHENOPTERIS LATIFOLIA.

Sph. latifolia, *Ad. Brongn. prodr.* 51. *Hist. des vég. foss.* 1. 205. *t.* 57. *f.* 1—5.

From the Bensham and Jarrow coal-mines where it is common.

A twice or thrice pinnated plant, the pinnæ of whose leaves vary very much in different specimens. They are generally, as represented in the plate, broad and blunt, with a heart-shaped outline, and consisting of about five rounded, nearly equal segments, which are divided almost down to the midrib. But occasionally they consist of seven segments, the lowermost of which are three-lobed and the upper confluent; in other cases they have almost constantly only three rounded segments, but in that case the middle segment is usually two-lobed ; and finally it sometimes happens that the segments are all run very much together, so as to destroy the deeply pinnatifid character of the pinnæ. This is so remarkably the case in Adolphe Brongniart's upper right-hand figure, that one could hardly avoid doubting whether it is really a portion of the same plant, if one did not know how very variable a species this is.

In all cases the veins diverge from the base of each segment, and spread into the parenchyma by a regular system of forking ; they never approach each other near enough to give the lobes a streaked appearance to the naked eye.

INDEX TO VOLUME II.

The Synonymes are printed in *Italics*.

THE END OF VOL. II.

NORMAN AND SKEEN, PRINTERS, MAIDEN LANE, COVENT GARDEN.

THE

FOSSIL FLORA

GREAT BRITAIN;

OR,

FIGURES AND DESCRIPTIONS

OF THE

VEGETABLE REMAINS FOUND IN A FOSSIL STATE

IN THIS COUNTRY.

BY

JOHN LINDLEY, Ph. D. F.R.S. G.S. &c.

PROFESSOR OF BOTANY IN THE UNIVERSITY OF LONDON;

AND

WILLIAM HUTTON, F.G.S. &c.

" Avant de donner un libre cours à notre imagination, il est essentiel de rassembler un plus grand nombre de faits incontestables, dont les conséquences puissent se déduire d'elles-mêmes."—*Sternberg.*

PART. I. OF VOLUME II.

LONDON:

JAMES RIDGWAY AND SONS, PICCADILLY.

1833—4.

Plate 80.

⅓ Natural Size

Publish'd by Ridgway & Sons, London, July 1853.

Plate 81

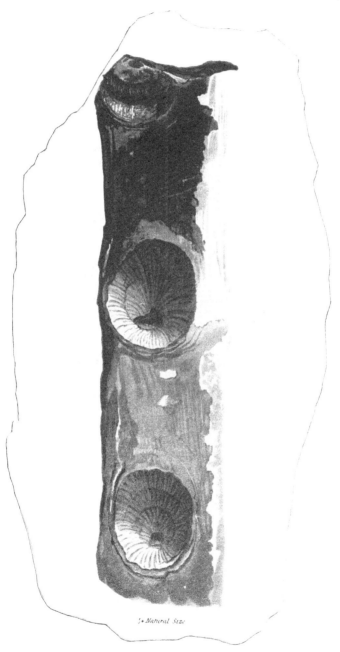

⅓ Natural Size

Published by Ridgway & Sons, London. July. 1833.

Plate 82

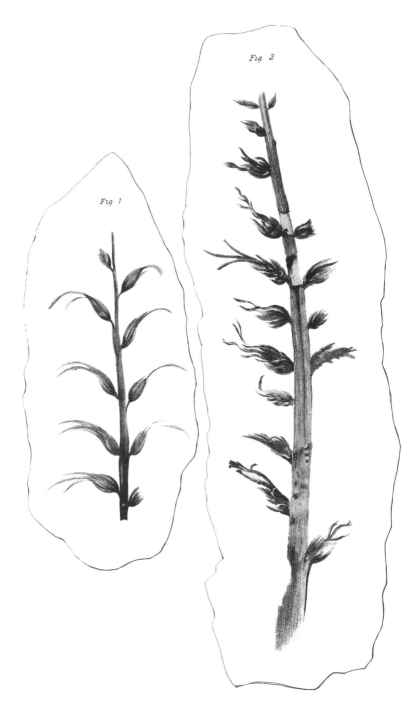

Fig 1

Fig 2

Published by Ridaway & Sons London July 1833

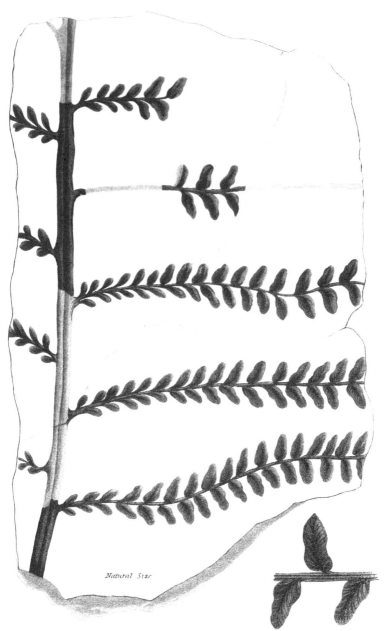

Natural Size

Magnified

W Wilkinson delt

Published by Ridaway & Sons London July.1833.

Plate 94

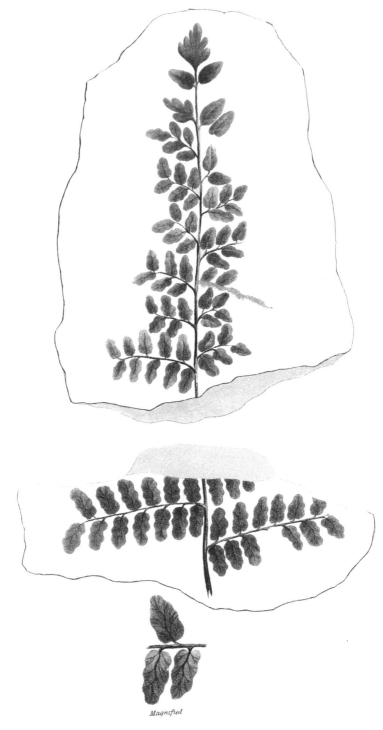

Magnified

Published by Ridgway & Sons, London, July, 1833.

Plate 85

Fig 1

Fig 2

A

B

Reversed

B

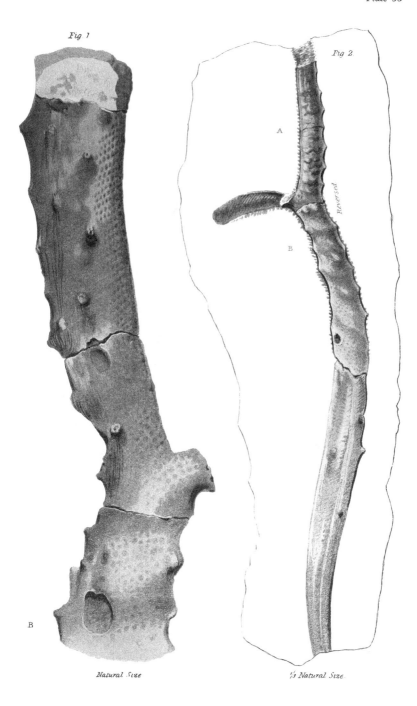

Natural Size

½ Natural Size

Published by Ridaway & Sons London July 1833

Plate 86

Fig 2

Fig 1

Natural Size

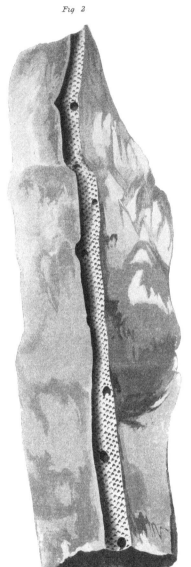

½ Natural Size

Published by Ridgway & Sons, London. July. 1833.

Plate 87

Fig 1

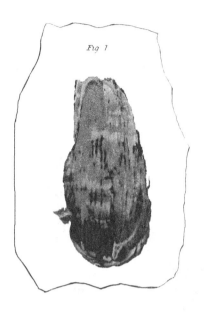

Fig 2

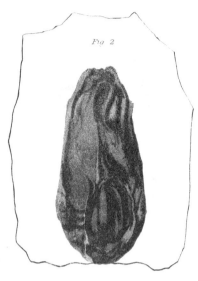

Fig 3

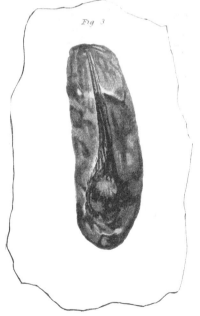

Fig 4

Natural Size

Published by Ridgway & Sons, London. Feb. 18.15

Plate 88.

Published by Ridgway & Sons, London, July, 1833.

Plate 89.

Natural Size

Published by Ridgway & Sons, London. July. 1833.

Plate 9

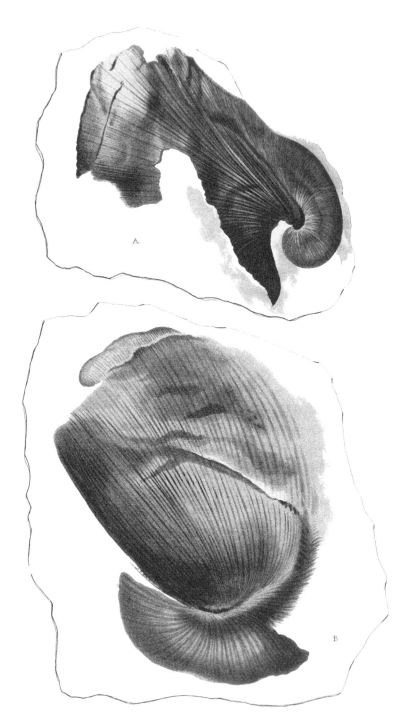

A

B

Published by Ridgway & Sons. London. Oct.r 1835.

Plate 91.

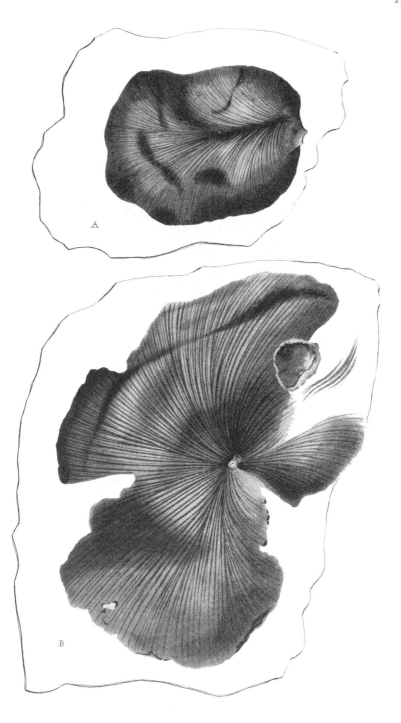

A

B

Published by Ridgway & Sons, London. Oct.r 1833.

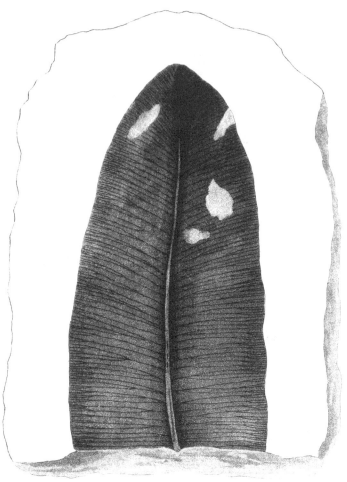

Natural Size

Magnified

Published by Ridgway & Sons, London. Oct.ʳ 1833.

Plate 93

Fig 2

Fig 7

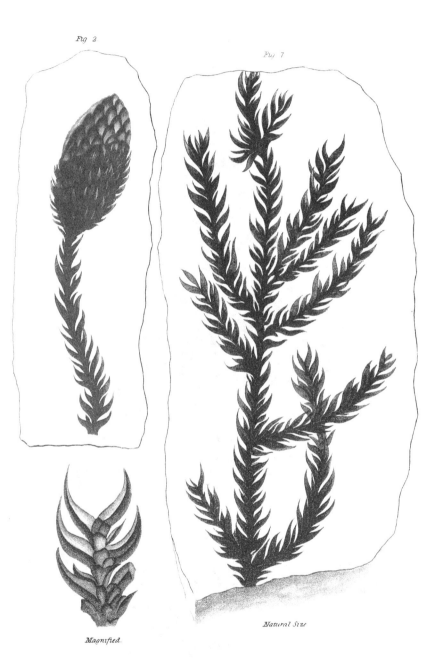

Magnified.

Natural Size

Published by Ridgway & Sons. London. Oct.ʳ 1833

Plate 94

Natural Size

Plate 95

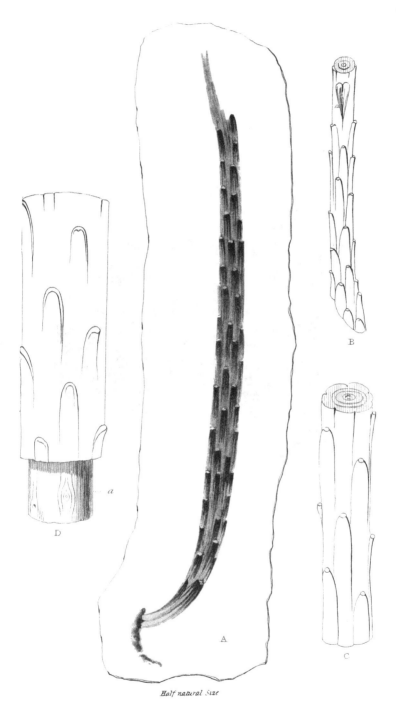

B

C

D

a

A

Half natural Size

Plate 96.

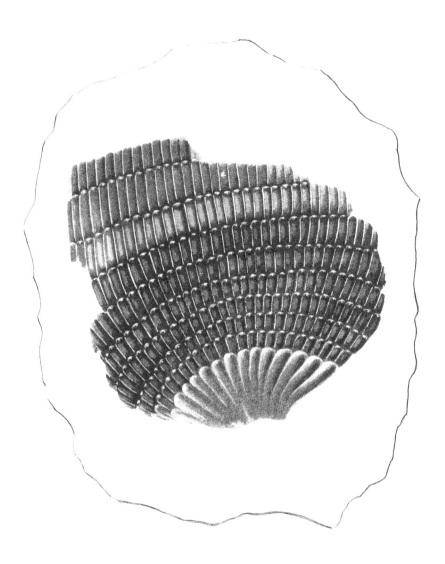

Published by Ridgway & Sons London Oct.r 1833.

Plate 97.

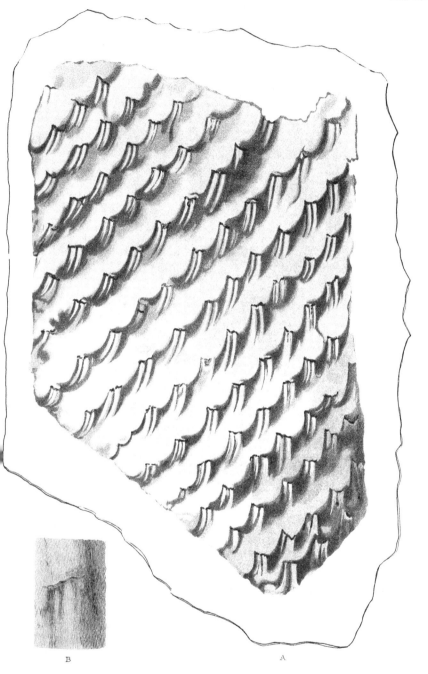

B   A

Published by Ridgway & Sons. London. Oct.r 1833.

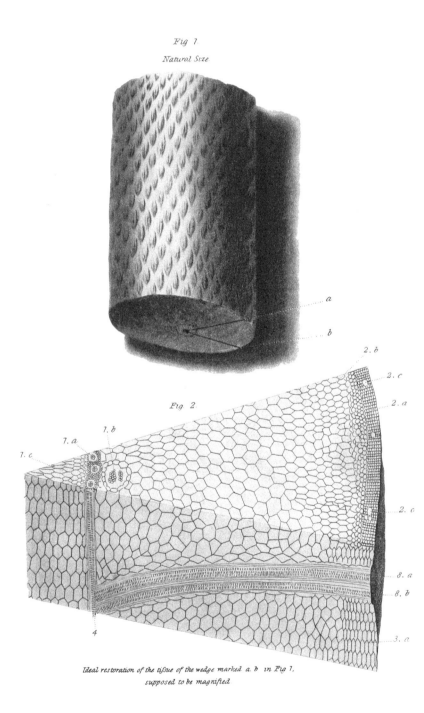

Plate 98

Fig 1.

Natural Size.

a

b

2. b
2. c
2. a

Fig 2.

1. b
1. a
1. c

2. c

8. a
8. b

4

3. a

Ideal restoration of the tissue of the wedge marked a. b in Fig 1,
supposed to be magnified

Plate 99.

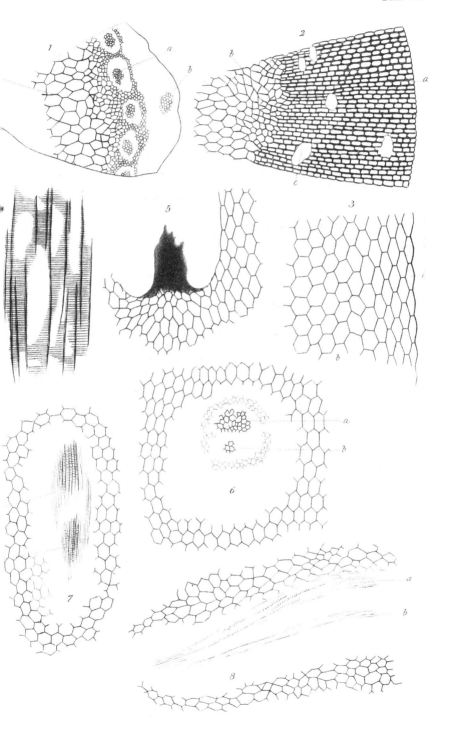

Plate 100.

'₂ Natural Size

Published by Ridgway & Sons, London, Jan.ʳ 1834.

Plate 101

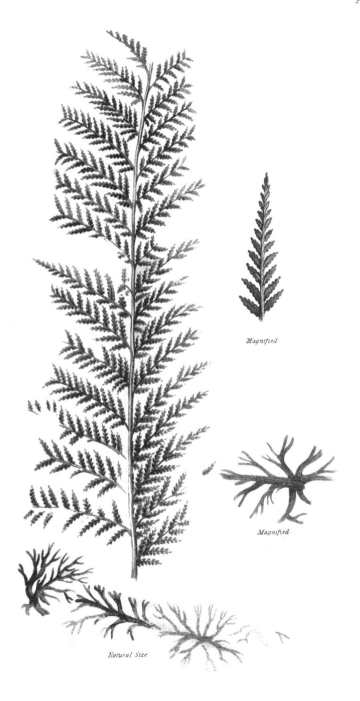

Magnified

Magnified.

Natural Size.

Published by Ridgway & Sons, London. July, 1833.

Natural Size.

Magnified

W.Williamson.del.

Published by Ridgway & Sons, London. Jan.ʸ 1834.

Plate 103.

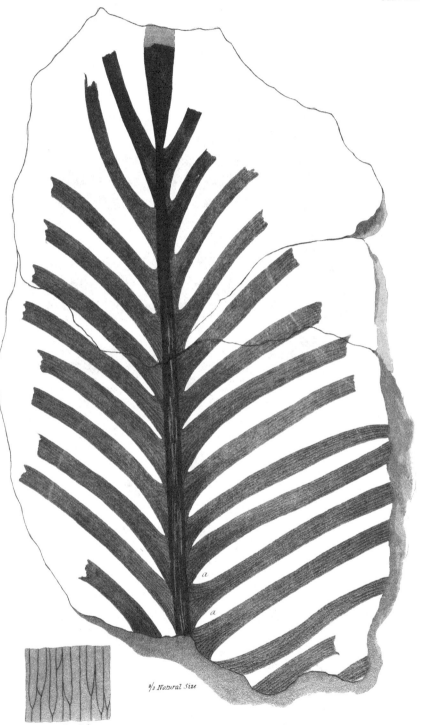

a

a

²⁄₃ *Natural Size*

Magnified

Published by Ridgway & Sons, London. July. 1833

Plate 104

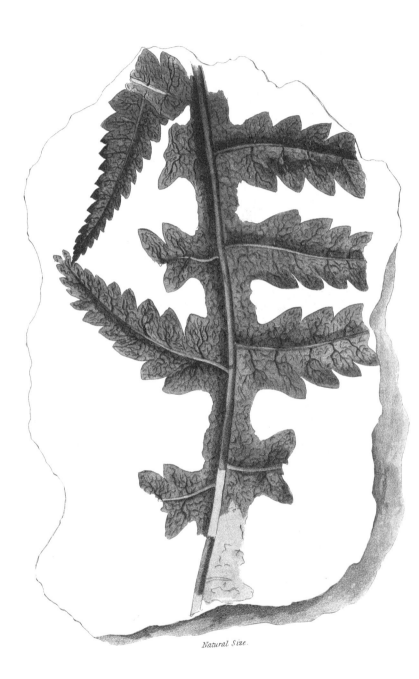

Natural Size.

Published by Ridgway & Sons, London, Jan.ʸ 1834.

Plate 105.

Magnified

Natural Size.

Magnified.

Published by Ridgway & Sons, London, Jan.ʸ 1834.

Plate 70

Natural Size

Published by Ridgway & Sons, London, Jan.ʸ 1834.

Plate 107

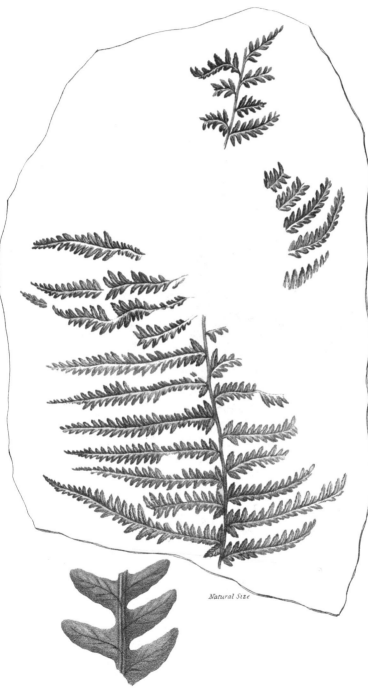

Natural Size

Magnified.

Published by Ridgway & Sons, London. Jan.ʸ 1834

Plate 108.

Natural Size.

Published by Ridgway & Sons, London, Jan.? 1834.

Plate 109.

Magnified.

Published by Ridgway & Sons. London. Jan.y 1834.

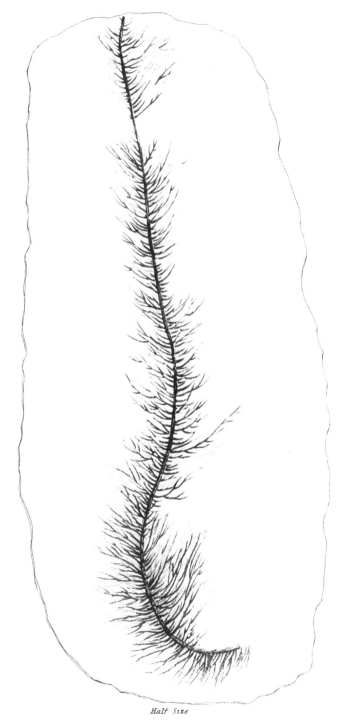

Half Size

Pubᵈ by Mefsʳˢ Ridgway. London. April 1834.

Pl. 111

Pl. 112

A

B

C

Publ. by Mess.rs Ridgway, London, April 1834

Pl 773

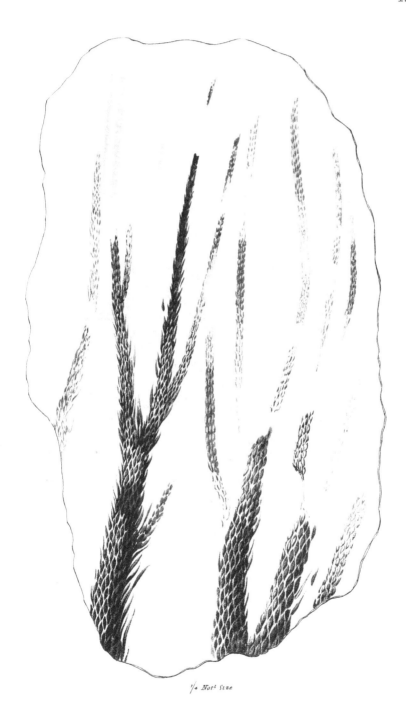

¹/₄ Nat¹ Size

Pub.ᵈ by Mᵉˢ.ʳˢ Ridgway London, April 1834

Pl. 11.

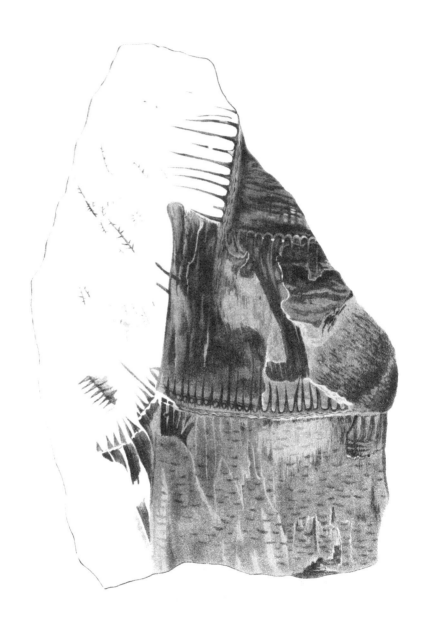

Pl. 115

Pl 116

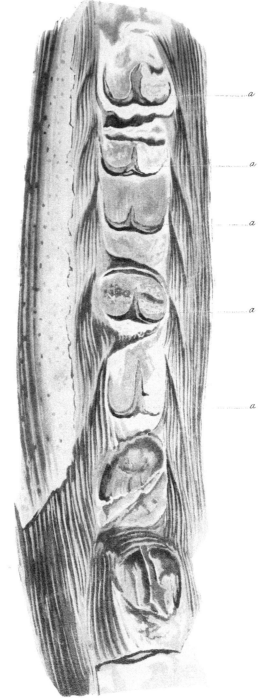

Pl. 117

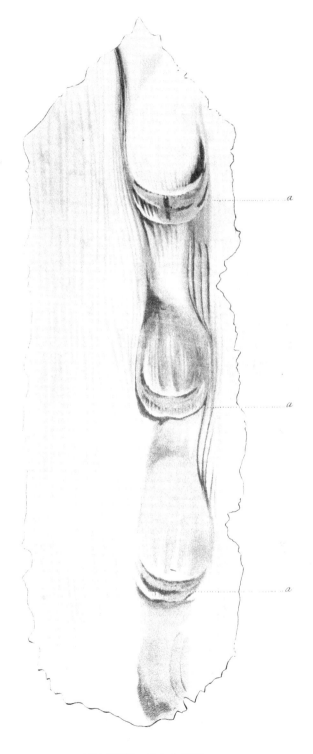

a

a

a

Pub. by Mess.rs Ridgway, London. Aprel 1834.

Jas.ᵗ Grar.   *Nat.ᵗ Grar.*

Pub.ᵈ by Mesᵗˢ Ridgway, London, April 1834.

The material originally positioned here is too large for reproduction in this reissue. A PDF can be downloaded from the web address given on page iv of this book, by clicking on 'Resources Available'.

Plate 119.

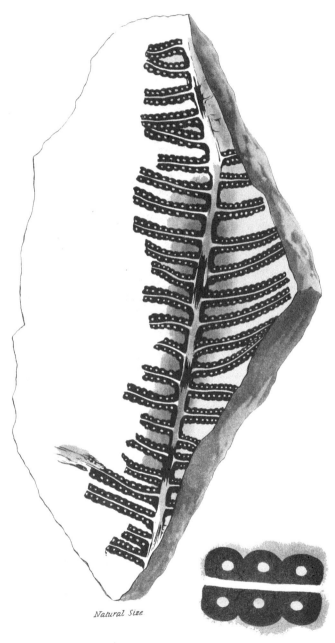

Natural Size

Magnified

Pub. by Mess.rs Ridgway. London, July 1834.

Plate 120

A

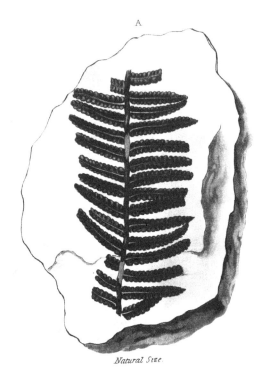

Natural Size.

C

Magnified

B

Horizontal view

Pub. by Mefs.ʳˢ Ridgway, London, July 1834.

Plate 121

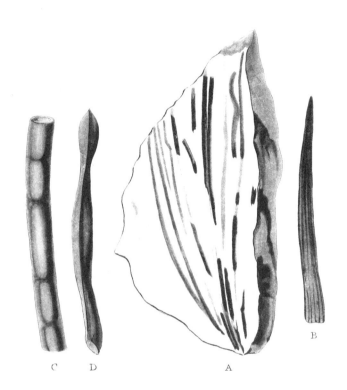

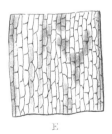

Plate 12

Plate 123.

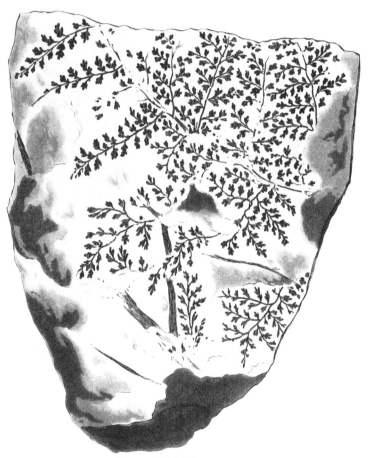

Natural Size.

Magnified

Plate 124.

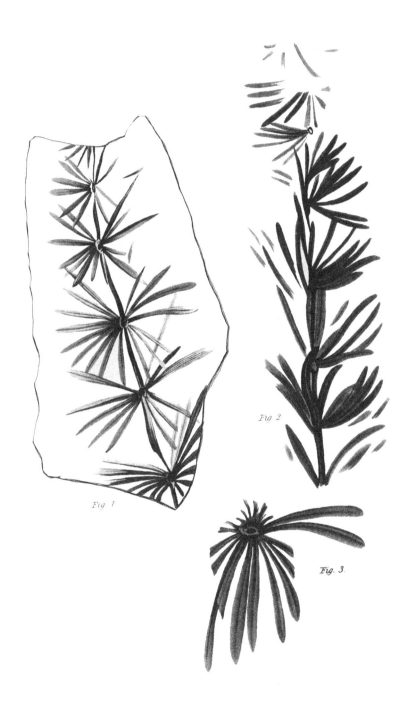

Fig. 1

Fig. 2

Fig. 3.

Plate 125

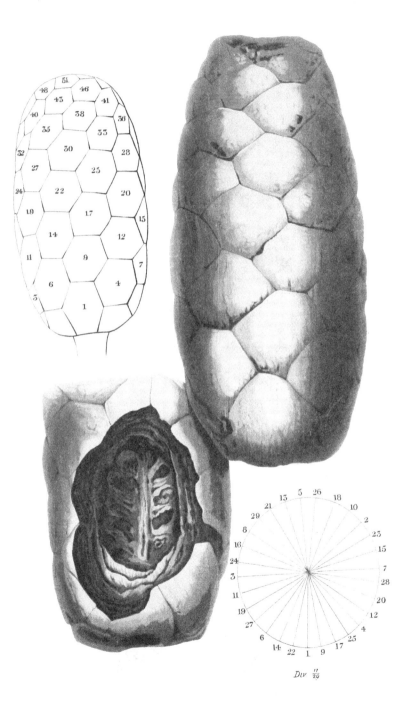

Div $\frac{11}{20}$

Plate 12

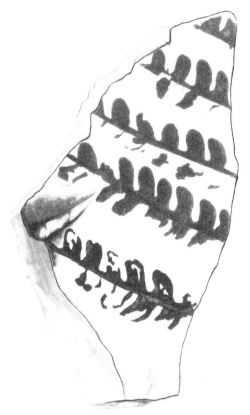

Natural Size

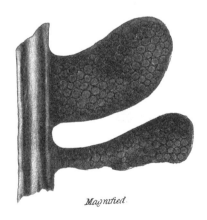

Magnified.

Plate 127

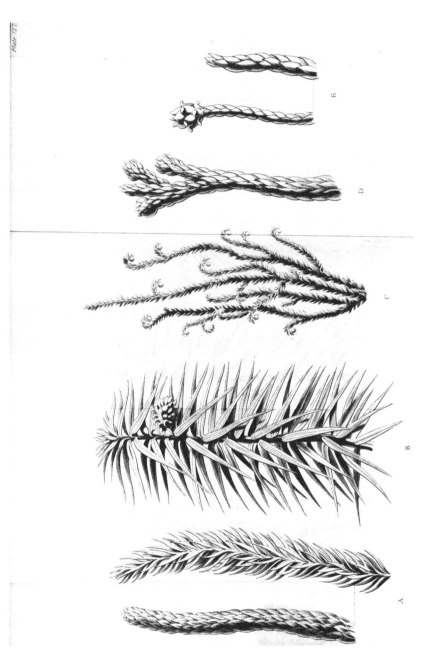

The material originally positioned here is too large for reproduction in this reissue. A PDF can be downloaded from the web address given on page iv of this book, by clicking on 'Resources Available'.

Plate 12.

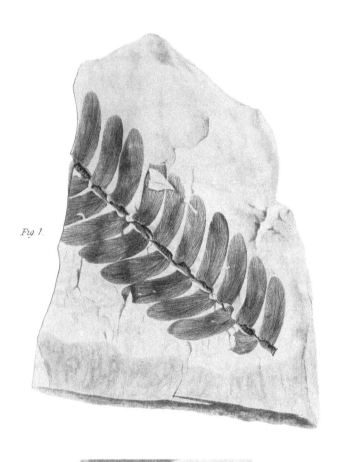

Fig. 1.

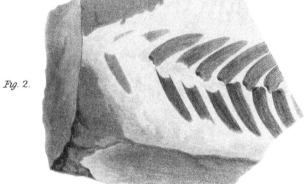

Fig. 2.

Natural Size.

Pub. by Mefs.rs Ridgway, London, Oct.r 1834.

Plate 129.

Fig. 1.

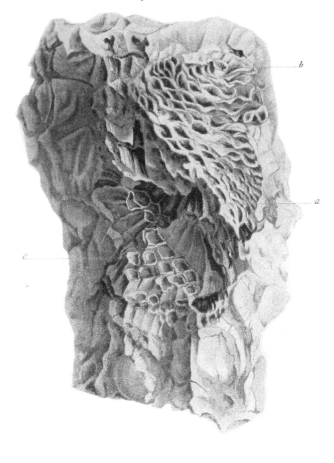

Fig. 2.

Pub. by Mess.rs Ridgway, London, Oct.r 1834.

Plate 130

Plate 131

Natural Size

Magnified

Natural Size

Pub. by Mess.ᵖˢ Ridgway London Oct.ʳ 1834

Plate 132

Natural Size

Plate 133

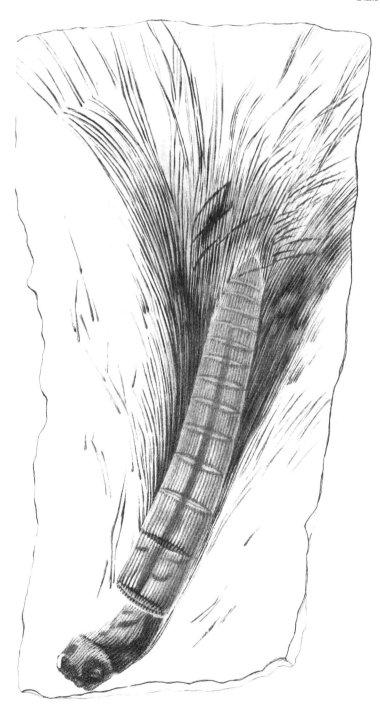

Pub.d by Mess.rs Ridgway, London, Oct.r 1834.

Plate 1.

Magnified

Natural Size

Pub. by Mess.rs Ridgway, London. Oct.r 1834.

Plate 135.

Natural Size

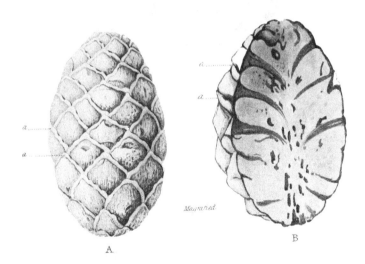

Magnified

A

B

Pub by Mess.rs Ridgway, London. Oct.r 1834.

Plate 136.

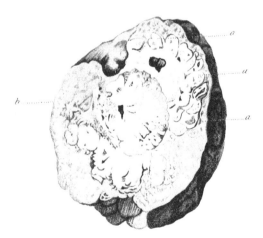

Pub. by Mess.rs Ridgway, London. Oct.r 1834.

Plate 137

1

2

Pub. by Mess.rs Ridgway, London. Oct.r 1834.

Plate 1.

Magnified

Natural Size

Pub. by Mes'rs Ridgway, London. Oct.' 1834.

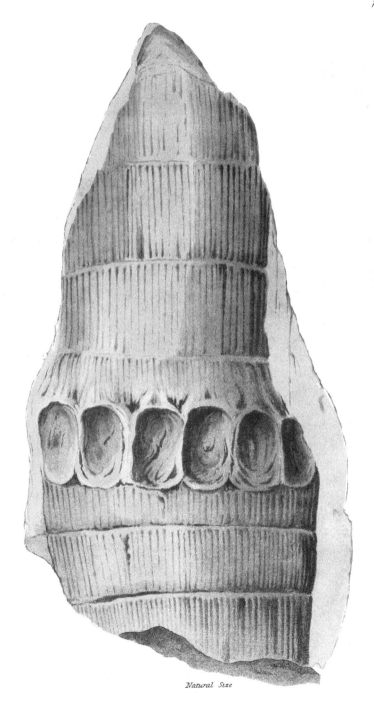

Natural Size

Pub. by Mes.rs Ridgway, London. Oct.r 1834.

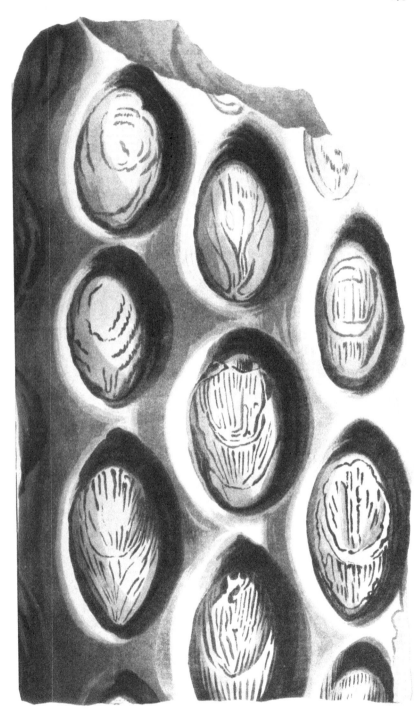

Pub. by Mes.rs Ridgway, London. Oct.r 1834.

Plate 141

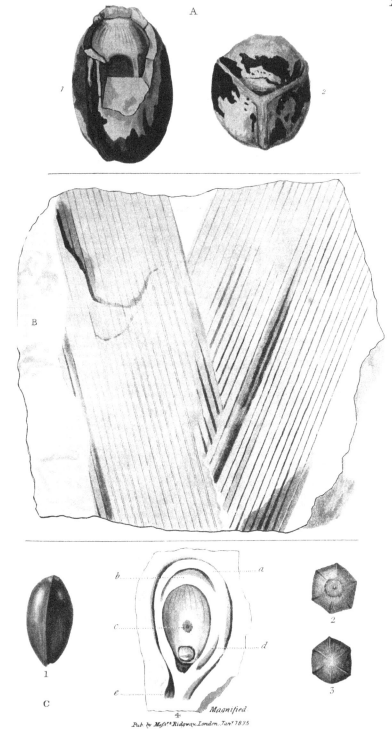

A

1 2

B

C 1 2 3 a b c d e 4 *Magnified*

Pub. by Mess.rs Ridgway. London. Jan.y 1835

Plate 143

Natural Size

Natural Size.

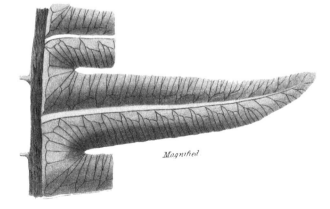

Magnified

Plate 145.

The material originally positioned here is too large for reproduction in this reissue. A PDF can be downloaded from the web address given on page iv of this book, by clicking on 'Resources Available'.

Plate 147

Magnified.

Pub. by Mess.rs Ridgway. London. April. 1835.

Plate 1

Magnified.

Pub. by Mess.rs Ridgway, London April. 1835.

Plate 149.

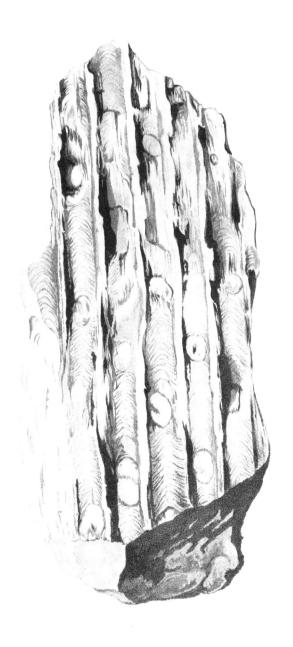

Pub. by Me'rs. Ridgway, London, April, 1835.

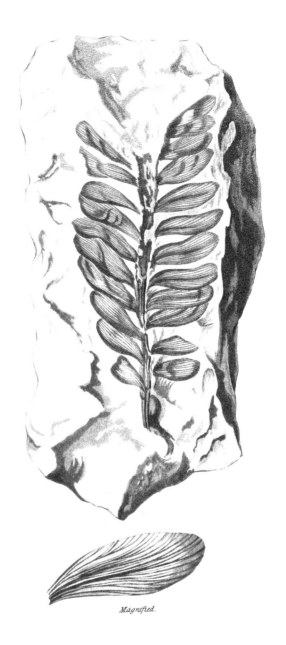

Magnified.

Pub. by Mess.ʳˢ Ridgway, London, April, 1835.

Plate 161.

Magnified.

Pub: by Mefs.ʳˢ Ridgway. London. April. 1835.

Plate 152.

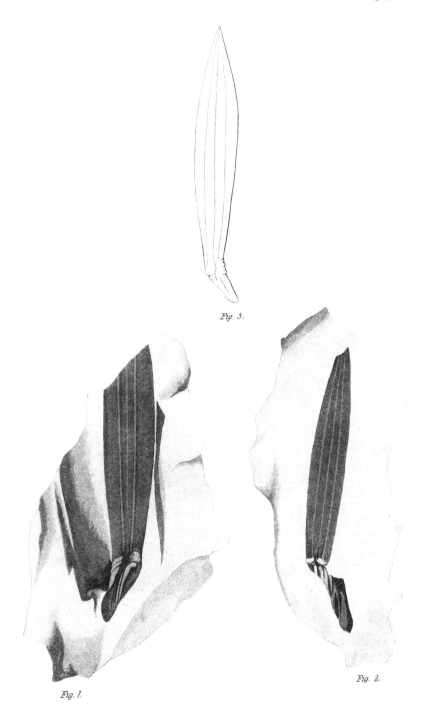

Fig. 3.

Fig. 1.

Fig. 2.

Pub. by Mesʳˢ Ridgway, London, April 1835.

Plate 153.

Magnified.

2

Magnified.

Pub. by Mefs.rs Ridgway. London. April. 1835.

Plate 154.

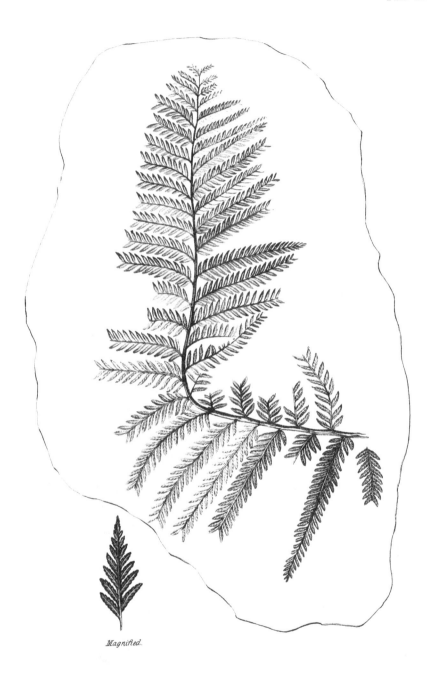

Magnified.

Pub. by Mefs.ʳˢRidgway. London. April 1835.

Plate 155.

Natural Size.

Magnified.

Pub. by Mefs.rs Ridgway. London, April. 1835.

Magnified.

Pub. by Mess.rs Ridgway, London. April, 1835.

Printed in the United States
By Bookmasters